Contents

	List of Figures and Tables	v
	List of Abbreviations	vii
	Foreword	xi
	Preface	xvii
	Acknowledgments	xix
	Introduction	xxi
Chapter 1	**Waging War**	1
	The Emergence of Limited War	1
	The Vietnam War Legacy	2
Chapter 2	**Making Decisions**	11
	Classic Theory in Decision Making	11
	Principles in Public Administration	19
	A Systems Approach?	24
Chapter 3	**The U.S. Military and the Art of War**	29
	Military Theorists	29
	Strategic Theory	30
	Military Doctrine	35
	Efficacy of the Principles of War	38
	The Principles of War	44
	Levels of War	58
Chapter 4	**National Security Policy Process**	67
	The United States' National Security Process	72
	The South African National Security Process	84
Chapter 5	**The Gulf War Case**	99
	Prelude to Invasion	99

	Desert Shield	103
	Desert Storm	112
	War Termination	124
Chapter 6	**The South African Case**	127
	Angola--The Country	127
	Prelude to Invasion	129
	South Africa and Angola: 1976-1984	134
	South Africa and Angola: 1985-1988	150
Chapter 7	**Application of the Principles of War**	169
	The Gulf War Case	170
	The Angolan War Case	178
	Comparing the Two Cases	187
	A Special Case: War Termination	190
Appendix	**Survey Questionnaire**	191
	Glossary	197
	Selected Bibliography	199
	Index	217

WAR AS AN INSTRUMENT OF POLICY

Past, Present, and Future

David V. Nowlin
Ronald J. Stupak

University Press of America,® Inc.
Lanham • New York • Oxford

Copyright © 1998
University Press of America,® Inc.
4720 Boston Way
Lanham, Maryland 20706

12 Hid's Copse Rd.
Cummor Hill, Oxford OX2 9JJ

All rights reserved
Printed in the United States of America
British Library Cataloging in Publication Information Available

Library of Congress Cataloging-in-Publication Data

Nowlin, David V.
War as an instrument of policy : past, present, and future / David V.
Nowlin and Ronald J. Stupak.
p.   cm.
Includes bibliographical references.
1. Military art and science. 2. International relations. 3. Angola—
History—South African Incursions, 1978-1990. 4. Persian Gulf
War, 1991. I. Stupak, Ronald J. II. Title.
U102.N624 1997   327.1'6—dc21   97-25301   CIP

ISBN 0-7618-0843-4 (cloth: alk. ppr.)
ISBN 0-7618-0844-2 (pbk: alk. ppr.)

∞™ The paper used in this publication meet the minimum
requirements of American National Standard for information
Sciences—Permanence of Paper for Printed Library Materials,
ANSI Z39.48—1984

Figures

2.1	The Systems Approach	24
2.2	Crisis Decision Making	25
2.3	Action Cycle	27
3.1	Levels of War	60
4.1	Framework for Grand Strategy	68
4.2	Department of Defense Chain of Command	77
4.3	Factors Influencing Decision Making	82
4.4	Representation of the State Government	86
4.5	Composition of the SADF	91
4.6	Racial Composition of the Full-Time Force	92
4.7	SADF Organizational Structure	93
5.1	The Coalition Dual Chain of Command	108
5.2	Sorties by Phase in the Air Campaign	115
5.3	Iraqi Resupply Movements	117
5.4	Iraqi Equipment Degradation in KTO Prior to G-Day	118
5.5	Arrival of U.S. Ground Forces	121
5.6	Iraqi Losses During the Gulf War	125
6.1	The African Continent	127
6.2	Map of Angola	128
6.3	Number of Terrorist Acts in South Africa—1976 to June 1988	142
6.4	Area of Operations	155
7.1	Gulf War Survey Mean and Standard Deviation	170
7.2	U.S. Chain of Command—CENTCOM AOR	173
7.3	Principles of War Applied in Angola	178
7.4	SADF Chain of Command	180
7.5	Principles of War Analysis—Fighters vs. Non-fighters	186
7.6	Comparing the Gulf War with Angola	187
7.7	Standard Deviations Compared	189

Tables

3.1	Principles of War	44
5.1	Coalition Order of Battle	109
5.2	Estimated Iraqi Order of Battle	110
6.1	Soviet Miliary Aid to Angola 1956-1987	135
6.4	Major SADF Operations Inside Angola 1980-1984	149

Abbreviations

AOR	Area of Responsibility
ANC	African National Congress
ARCENT	Army Component Commander, Central Command
ASUW	Antisurface Warfare
CENTCOM	Central Command
CIA	Central Intelligence Agency
CINC	Commander in Chief
CINCCENT	Commander in Chief, Central Command
CJCS	Chairman of the Joint Chiefs of Staff
COIN	Counterinsurgency
CRAF	Civil Reserve Air Fleet
CSI	Chief of Staff, Intelligence
FAPLA	Popular Forces for the Liberation of Angola (Military wing of the MPLA)
FNLA	Frente Nacional de Libertaçao de Angola (National Front for the Liberation of Angola)
GOC	General Officer Commanding
ICAF	Industrial College of the Armed Forces
JCS	Joint Chiefs of Staff
JFACC	Joint Force Air Component Commander
KTO	Kuwaiti Theater of Operations
LOC	Line of Communications

MCM	Mine Countermeasures
MEF	Marine Expeditionary Force
MK	Umkhonto we Sizwe (armed wing of the ANC)
MPLA	Movimento Popular de Libertaçao de Angola (Popular Movement for the Liberation of Angola)
NBC	Nuclear, Biological, and Chemical
NCA	National Command Authority
NDU	National Defense University
NIS	National Intelligence Service
NP	National Party
NSC	National Security College
NWC	National War College
OAU	Organization of African Unity
OI	Operations Instruction
PA	Public Administration
PLAN	People's Liberation Army of Namibia
PMC	Political Military Council
RGFC	Republican Guard Forces Command
ROE	Rules of Engagement
RSA	Republic of South Africa
SAAF	South African Air Force
SAC	Strategic Air Command

SACP	South African Communist Party
SADF	South African Defence Force
SAI	South African Infantry
SAM	Surface to Air Missile
SAP	South African Police
SSC	State Security Council
SWA	South West Africa
SWAPO	South West Africa People's Organization
UNITA	União Nacional para a Independéncia Total de Angola (National Union for the Total Independence of Angola)
UNSC	United Nations Security Council

Foreword

War as an Instrument of Policy: Past, Present, and Future is a well-considered attempt to provide a rational, structured approach to the question of whether and how to use military power. The authors' thesis is that application of the classic principles of war can not only lead to success on the battlefield, but also provide the basis for good decision making at the highest levels of government where the military and civilian leaders come together. They do this by analyzing the principles of war and applying them to the Angola campaign of 1987 and Operation Desert Storm in 1990.

Principles of war are nothing more than "lessons learned." A problem with any attempt to produce a useful and valid compendium of "lessons learned" is correlation: how similar are the situations; how similar are the tools to be used in dealing with them. The problem is compounded by the human tendency to see things through the lenses of our experience. This is no less true when attempting to assemble the "lessons learned" of war. The wars of today are like and yet unlike the wars of yesterday; certainly, many of the tools of war--weapons, methods of communication, and the like--are different. The classic writers on war, Baron Antoine Jomini and Karl von Clausewitz, were products of the Napoleonic era, having served in Napoleon's army in a number of his most famous campaigns. Another, Captain B.H. Liddell Hart, was a product of WWI, and his extensive writings on all wars were likely to have been influenced by that experience. Given all this, is it still possible to speak of a set of immutable *principles* of war from the past that will serve us well into the future? The authors make a persuasive case that it is not only possible but necessary to do so.

My own acquaintance with the classic principles of war came almost immediately upon my arrival as a cadet at West Point in the summer of 1954, one year after the Korean cease-fire of July 27, 1953. The "Nine

Principles of War" were tenets of faith, to be memorized and put to use. I remember studying the great campaigns of history, from Miltiades's tactics at the Battle of Marathon some 500 years B.C., to the Battle of the Bulge in WWII, with the maps in one hand and the Nine Principles of War in the other. Later, during my year in Vietnam, which began just after the Tet offensive of 1968, I marveled at how consistently those principles had been violated.

What the authors have done in this book is stimulate me to push their ideas even further. Thus what follows is a series of "yes, ands" meant to acknowledge their contribution and cry for extensions.

The first "yes, and" has to do with the "everybody is like us" trap. We tend to view war from the standpoint of a great democracy whose record of warfare for a long time has been defensive or protective rather than offensive or aggressive. All of the democracies tend to negotiate in good faith first (not just present nonnegotiable demands), leaving war as a last, and overwhelmingly avoided, resort. But many nations of the world, the ones we are likely to have the most difficulty with, are not democracies. For nondemocratic nations, war is less an extension of "national policy by other means" than the ultimate extension of ego. Our forefathers were wise to set up a government of distributed egos rather than allow the accumulation of too much ego in one place. Perhaps we need some subset of the classic war principles, or even a new set of principles, to deal with "Great Ego" wars. Ego tends never to go away, leaving problems of war termination and ensuing "state-of-peace" tensions to fester for a number of administrations, even when we win. When we lose, the opportunity for infinite grief goes up infinitely.

Another "yes, and" has to do with setting *bounds* on the principles. As I see it, each of the principles has a "too little/too much" aspect that deserves some analysis. For example, too little Unity of Command leads to anarchy and chaos. Too much can lead to excessive risk-taking and lack of consideration of alternatives. How does one achieve an optimum balance? How does one know one has found it?

In the modern age, one cannot overlook the double-edged sword of technology. A report written after the Yom Kippur War of October 1973, based upon interviews with Israeli government and military officials, outlined the Israelis' worst deficiencies. Surprisingly (to me, at least), at the top of the list was "not knowing where our own troops were." Second was "not knowing where the enemy was." Technology can fix the first one. At any Eddie Bauer store it is possible to buy a hand-held receiver for $400 or so, which, by tuning in several Global Positioning Satellites, can result in position location within less than 30 meters. Every soldier (or at least, every squad leader) can have one. It would then be possible for commanders to have the same information in near real time.

But would commanders have the discipline not to indulge the human passion for overcontrol? As noted by the authors, battalion commanders in Vietnam tried to direct battlefield operations from a helicopter; worse, President Johnson tried to direct the Vietnam War from the basement of the White House. Contrast this with the story of the soldiers in the Grenada operation who called in support requests from a phone booth because their radios would not work with those of the other services. The media, of course, seized the opportunity to blast the Defense Department (rightly) for buying communications systems that lacked interoperability. But to me, the story illustrated one of our nation's greatest strengths: the ability and willingness of our troops to use initiative. We do not want to lose this in a wrong-headed effort to perfect unity of command. Too much information (Martin Van Creveld calls this "information pathology"[1]) leads to too much control. Perhaps the ideal is what Tom Peters calls "loose-tight" management.

Another "yes, and" would entail a fuller treatment of the principle of mass, particularly as regards logistics. A "mass" of troops without the resources to support them is almost useless. A delicious quote of unknown origin goes as follows (paraphrased): "If you study the great military campaigns of history, you will find that the losers studied tactics; the winners studied logistics." Napoleon and Hitler outran their supply lines with devastating results. The Normandy invasion was not a triumph of tactics so much as it was a triumph of logistical support. The real heroes of Normandy, it could be argued, were the builders of the assault craft, gliders, portable bridges, nonperishable foodstuffs, parachutes large enough to carry trucks and jeeps, the radios that allowed people to coordinate their efforts in unfamiliar terrain, the training materials, and all the other elements that contributed so mightily to the effort. As recorded by Lynn Montross, "the Allies were able to land 326,547 men in the first six days, in addition to 54,186 vehicles and 104,428 tons of stores."[2]

We have been lucky or prescient to have had so much time to prepare for several of our major war efforts. We may have been spoiled. In WWI, although the first American units arrived in France in late 1917, they were not committed in force for nearly six months. When committed, they were well trained, equipped, and ready to go. Eisenhower was appointed to command the Overlord operation (Normandy invasion) in January 1943, five months before the invasion, and preparations had begun before that. As the authors report, in Desert Storm, the Iraqis completed the occupation of Kuwait in the 36 hours

---

[1] Van Creveld, Martin, *Command in War* (Cambridge, MA: Harvard University Press, 1985), 249.

[2] Montross, Lynn, *War through the Ages* (New York: Harper & Brothers, Revised and Enlarged Third Edition, 1960), 927.

preceding August 3, 1990. The coalition forces, however, did not attack until January 17, 1991, nearly five and one-half months later, after a massive infusion of troops and supplies and time to spend learning how to operate in the desert environment.

In Korea, however, the story was different. When the North Korean Army attacked the Republic of Korea (ROK) Army on June 25, 1950, "'the best damn army outside the United States' had no tanks, no medium artillery, no 4.2-inch mortars, no recoilless rifles. They had no spare parts for their transport. They had not even one combat aircraft."[3] President Truman immediately dispatched two divisions from Japan. Because of the efficiency of the American drawdown after WWII, however, the finest fighting force the world had ever seen was down to only ten divisions and nine regimental combat teams, all of which were at reduced strength, none of which having its proper wartime quota of weapons.[4] Within seven weeks, the original divisions plus other UN reinforcements had been pushed into a small perimeter around the port of Pusan, where a gallant defense finally allowed the "Arsenal of Democracy" to come to the rescue. It was close, very close.

This brings me to the final point in my list of worthy ideas to be pushed further. More study is needed to determine a set of principles to be followed in the event the classic principles *cannot* be followed. Clearly, in the modern world, an ideal Unity of Command may not be achievable. Could a Desert Storm-type coalition be assembled again? How should war be conducted when the best one can do is a hodgepodge of command? Should war even be considered without it? Are certain combinations of principles powerful enough to offset the inability to observe others? The principle of "security" was violated in Vietnam, but that was partly because no one was sure who or where the enemy was. Sometimes it is impossible to achieve surprise. Perhaps achieving the "offensive" is impractical, as in the Lebanon situation. But we live in a nasty world, and we need to take action sometimes when all the choices are terrible. Perhaps treatment of the principles of war along these lines would give us some fresh ideas for how to cope with it better.

The authors have written an excellent treatise that makes a worthy contribution to the field of public endeavor. Their creative blend of the principles of war with tenets of public administration provides an innovative approach to decision making along, as they put it, the "seam" of civilian/military discourse and action. I would certainly

---

[3] Fehrenbach, T.R., *This Kind of War* (New York: Macmillan, 1963), 17. The internal quote is by Brig. Gen. William L. Roberts from a *Time* magazine article.
[4] Ibid.

recommend this book to the military war colleges. When I attended the Air War College in 1973-74, I complained that we discussed "war" very little and "air" even less. The big thing then was "management." This book would have been a valuable addition to the curriculum. I would also recommend it to the administrators, both career and politically appointed, of the Department of Defense. The holistic view taken in the book would give them a better perspective on their real jobs. Finally, I would recommend it to the students and faculty of the Defense Systems Management College (of which I was once Commandant). Those in charge of the procurement of weapons systems should have a better understanding of their uses and limitations.

Do I really need to quote George Santayana on the lessons of history?

Charles P. Cabell, Jr., DPA
Brig. Gen. USAF, (Ret.)

Preface

As abhorrent as it may be, warfare has routinely been chosen as a preferred method of dealing with discord among states. The use of military force as an instrument of foreign policy, however, introduces a complex set of issues into the decision-making process of a nation's executive leadership. This book examines this decision-making process, and the relationship between civilian and military decision makers in particular, from both an academic and functional perspective using two case studies -- the United States in the Persian Gulf War and South Africa in the Angolan War -- to form a structural framework for the inquiry.

The inquiry itself focuses on the disciplines of decision-making theory, executive and military leadership, national security processes, and military strategy and doctrine. It asks the question: How can the principles of war be transposed from framing the decision-making process on or near the battlefield to being used as an effective approach to decision making at the highest levels of government during a crisis situation? In exploring the answer, it suggests ways to facilitate and improve the decision-making process.

In fact, research conducted using the Gulf and Angolan Wars to evaluate the practical application of the principles of war indicates that there is a positive relationship between proper use of the principles and successful results stemming from military action. If the principles of war are to be used as a design for decision making, however, judicious employment should be exercised to prevent their use as an inflexible checklist, the study concludes.

But success on the battlefield does not necessarily achieve desired political results. War termination and the ensuing postwar state of peace must be considered concurrently with military concerns when initially responding to a crisis situation. The overall political objective can easily be overlooked when making critical decisions during the "heat of battle." History is replete with such occurrences.

Acknowledgments

We have been extremely fortunate in receiving valuable assistance, sage advice, undying support, and exemplary feedback from a host of students, scholars, colleagues, family, and friends. We would like to express our gratitude to all who made it possible to complete this book. We are indebted to the following people and organizations and the roles they played in nurturing this effort:

- Special recognition and thanks must go to Colonel Jim Toth (USMC, Ret.) who gave us many excellent ideas to pursue in the initial stages of the research.
- A former South African military officer, Brigadier Willem Van de Waals (Ret.), contributed indispensable assistance in the South African case by identifying invaluable sources for research and lending his local expertise for preserving accuracy.
- A special mention must be made about the South African Defence Force. Its assistance in identifying knowledgeable personnel and distributing survey forms was essential to the data collection phase of the effort.
- We are also indebted to all those military personnel and other professionals who gave of their valuable time to fill out the extensive, time-consuming questionnaire. Their candid and thorough answers contributed immeasurably to advancing the database of this inquiry.
- We owe a great debt to all the numerous reviewers and editors who helped to make each draft more readable, integrated, and sophisticated.
- Finally, we wish to sincerely thank our wives, significant others, and confidants for their patience over the two years we were consumed by this endeavor. Without their willingness to

make numerous sacrifices throughout the exercise, this would never have come to fruition.

Introduction

On 5 August 1987, a South African Defence Force (SADF) military convoy located in South West Africa (now Namibia) crossed over the border into Angola. This was one of many military incursions into Angola that had been initiated by the South African government over the previous 13 years, with the purpose of influencing Angolan political and military affairs in a manner favorable to South African internal and strategic interests.

Three years later and a continent away, on 2 August 1990, Iraqi military forces invaded the oil-rich country of Kuwait and threatened to do the same to the strategically important kingdom of Saudi Arabia. The United States and its allies responded immediately by first limiting the Iraqi offensive through the use of political and economic actions in coordination with defensive military operations, and then, when political policy options failed, liberating Kuwait through the use of offensive military force.

Both the United States and the South African governments felt a need to resort to military force to defend and/or advance national interests after more traditional political endeavors had, from their perspective, failed. In the United States' case, the requirement to employ military force was apparent. In the South African case, the need to mobilize the SADF was more ambiguous.

While nations have historically used war to further their objectives, the application of military force as an instrument of policy has often been misunderstood. The United States in particular has grappled with this issue since the end of World War II. No longer is the use of armed force as an instrument of policy as cut-and-dry as when national survival was at stake. Under the threat of nuclear annihilation, wars have become more limited in nature while the pursuit of national objectives and goals has become more complex. This drive for national interests now, more than ever before, impacts other nations within the so-called "global village."

One further consideration of using force is the consequence resulting from military operations. Should the decision be made to declare war or use force short of war to advance or defend national interests, a determination must be made on how to conduct operations so as to terminate the conflict in terms most favorable to the country's goals and well-being. In the end, an understanding of conflict termination is just as important as how a nation prosecutes armed intervention in the first place.

Coming Together at the Seam

One of the most difficult decisions a President may face during his term in office is whether or not to use military force to further the interests of the United States. Although the country has had only five declared wars in its 200-year history, the decision to use, or threaten the use of, military force has been more extensive than one might believe. A study conducted by Barry M. Blechman and Stephen S. Kaplan in 1975 revealed that the President was called upon to employ armed forces for political purposes 215 times between 1946 and 1975.[1]

A President's decision to use military force is not a simple yes/no decision. It is fraught with considerations concerning the proper employment of force and the potential consequences of doing so. For instance, the President has to consider the political counsel of his advisors along with the expectations of various interest groups, including public opinion. He also must seek the advice of his primary military advisor, the Chairman of the Joint Chiefs of Staff (CJCS); something he may be uncomfortable with or unwilling to do. How much should he trust in his own judgment, and how much should he rely on the military establishment to lead a U.S. response in redressing a critical foreign policy issue?

A seam forms at an important juncture in the decision-making process of our federal government. It is a seam where two cultures—military and civilian—come together in the hierarchy of the executive branch, often harmoniously, but sometimes contentiously. This seam forms a pivotal point where critical decisions must be made—whether or not to use military force, the establishment of goals and objectives,

---

[1] Barry M. Blechman and Stephen S. Kaplan, *Force Without War* (Washington, DC: The Brookings Institute, 1978), 16.

plans on how to prosecute war if required, and how to successfully terminate the conflict.

The use of war as an instrument of policy can be critical to the survival of the nation, but the decision-making structures used to determine if and how to manage the process are fraught with obstacles that may impede the application of even satisfactory leadership and management. Proper use of the principles of war can provide a rational approach to systematizing the process in order to minimize the unknowns.

Chapter 1

Waging War

> War is not merely an act of policy but a true political instrument, a continuation of political intercourse, carried on with other means . . . The political object is the goal, war is the means of reaching it, and means can never be considered in isolation from their purpose.[1]

The Emergence of Limited War

The concept of war fighting changed dramatically for "first world" countries after World War II. Gone were the days when military and industrial powers were required to mobilize their entire populations and economies when faced with the need to use military force. The advent of the Cold War and the introduction of nuclear weapons made the notion of total war in the conventional sense virtually obsolete. The concept of limited warfare crept into the lexicon of military theorists. Revolutions incorporating insurgencies and guerrilla-style warfare became the norm, but the dominant military powers were unable to cope.

Mao won China. The Viet Minh forced France out of Indochina. The Huks caused the United States concern in the Philippines. Castro prevailed in Cuba. Numerous countries in Latin and South America became embroiled in indigenous struggles supported by outside assistance. African countries embraced insurgent warfare to overthrow colonial rule. From Greece to Afghanistan to Burma to Peru and Angola, the nations of the first world were forced to deal with an

---

[1] Carl Von Clausewitz, *On War*, edited and translated by Michael Howard and Peter Paret (Princeton, NJ: Princeton University Press, 1976), 87.

unfamiliar approach to conflict; and they were not prepared to meet the challenge.

The United States created its own problems when confronted with the notion of limited war. During the Korean War the U.S. military was not able to view warfare without the objective being total victory. Yet, at the same time, U.S. leadership was afraid of the war escalating into a nuclear conflict with the Soviet Union, so the objectives of using military force and the aims of policy were never seen as compatible. "Still thinking in terms of total victories or total defeats, after the winter of 1950-51 the United States thought that stalemate was the only alternative to total war, because Americans assumed that Russia would not tolerate a successful American initiative," explained Russell Weigley in *The American Way of War*.[2]

Decisions made during the Korean War were a harbinger of the way American military force would be used to support foreign policy during the decades of the 1950s, 1960s, and 1970s. In Korea, and then again in Vietnam, military advisors were often frustrated by the tight civilian control over military operations, noting that, in many cases, strategic and operational decisions were made by the civilian leadership at the expense of contravening advice provided by their military advisors. Many military leaders believed this prevented a victory in Korea and led to defeat in Vietnam.

The civilian leadership, on the other hand, was more concerned with the two wars escalating into a superpower confrontation between the United States and the Soviet Union, complete with the exchange of nuclear weapons. This presumption, along with several other doctrinal concepts lost to both the political and military leadership, led directly to America's most ignominious application of military power in the history of this country and the failure to win in Vietnam.

The Vietnam War Legacy

Even 20 years after the fall of Saigon to North Vietnam, issues surrounding the failure of the United States to defend South Vietnam remain emotional and divisive both to a population who suffered through the long, drawn-out conflict and to the military personnel who dedicated their careers and lives in an honest attempt to win the war. The Persian Gulf War played a significant role in helping to heal the psychological wounds of the nation; but what we as a nation and

---

[2] Russell F. Weigley, *The American Way of War* (New York: Macmillan Publishing Co., 1973; Bloomington, IN: Indiana University Press, 1977), 415.

military establishment must never forget are the so-called lessons learned resulting from our failure. Retaining an awareness of how and why the employment of basic military strategy and doctrine was allowed to be neglected during the Vietnam War is an important part of the experience.

While the debate continues over why the United States failed in Vietnam, the disregard of fundamental tenets of military strategy, particularly at the national decision-making level, must be considered a major contributing factor. In fact, the way the Vietnam War was conducted from the very beginning foreordained a no-win strategic policy. Herbert Schandler described the process that evolved in an effort to attain American war aims:

> Gradual escalation was . . . the strategy chosen for achieving United States objectives. Domestic politics dictated the minimum necessary disruption in American life. But with each passing year of war, the domestic political position of the President grew weaker. Optimism without results produced, in time, a credibility gap.[3]

In other words, President Lyndon Johnson's policy objectives focused on using the least amount of military force to prevent a South Vietnamese defeat at the hands of the North Vietnamese. This policy eventually led to a loss in direction toward the real war objectives and the correct application of political recourse and military force to attain them.

The use of military force was initially based on the United States' policy objective of allowing South Vietnam to remain free to determine its own future without interference from external sources. Although this was a laudable and strategically appropriate objective, supporting political and military objectives were not defined, resources were not put in place, and proper execution was not effected to ensure this objective was achieved. In short, the situation was contrary to military doctrine on how to fight a war, as Schandler explained:

> The objective would not be to "win," either in North or South Vietnam, but rather to convince the North Vietnamese (and their Soviet and Chinese sponsors) that the cost of continuing the war in South Vietnam would be, over time, prohibitive to them and they could not succeed.
>
> In actuality, however, there was no clear conception in Washington as to when this elusive psychological goal would be achieved.

---

[3] Herbert Y. Schandler, *The Unmaking of a President, Lyndon Johnson and Vietnam* (Princeton, NJ: Princeton University Press, 1977), 335.

The President's strategy, then, was defensive in nature and, in effect, left the decision as to when to end the war in the hands of the North Vietnamese.[4]

The United States had, in fact, lost the initiative on how to prosecute the war. Ironically, this was the same approach employed by the French in their losing effort to defeat Ho Chi Minh in Indochina during the early 1950s.

Lack of resources to fight the war centered primarily around decisions on how many troops should be placed in the field. The constraining parameter was manpower in the regular force since the President drew the line at mobilizing the reserves.

As the war developed, the debate within the administration concerning the level of American effort in South Vietnam, in fact, came to revolve around this one crucial issue of mobilization. When the President searched for the elusive point at which the political costs of the effort in Vietnam would become unacceptable to the American people, he always settled upon mobilization, that point at which reservists would have to be called up to provide enough manpower to support the war.

This domestic constraint, with all its political and social implications, *not any argument concerning long-range military strategy*, appears to have dictated American war policy.[5]

Domestic political considerations notwithstanding, there were very real, legitimate concerns inhibiting the unrestrained use of military force to defeat the North Vietnamese: the potential escalation into a nuclear conflict with the Soviet Union and the possible intervention of the Chinese, à la Korea. Whether these fears were justified may never be reconciled; but the important point to consider is that the concern over escalation defined U.S. strategic policy, became the underlying principle for a no-win strategy of gradual escalation, and constrained the proper use of military force in the conduct of the war. According to Brigadier General David Palmer in his analysis of the Vietnam War:

> The Johnson Administration had already barricaded the one sure route to victory, to take the strategic offensive against the source of the war. Memories of Mao Tse-tung's reaction when North Korea was overrun by United Nations troops in 1950 haunted the White House.

---

[4] Herbert Y. Schandler, "America and Vietnam: The Failure of Strategy, 1964-67," in *Vietnam as History, Ten Years After the Paris Peace Accords*, ed. Peter Braestrup (Washington, DC: University Press of America, 1984), 24.

[5] Ibid., 25.

America's fear of war with Red China protected North Vietnam from invasion more surely than any instrument of war Hanoi could have fielded.[6]

Colonel Harry Summers (U.S. Army, Ret.) took great issue with the Administration's approach to fighting the war in his critically acclaimed (but not universally accepted) assessment, *On Strategy, A Critical Analysis of the Vietnam War*, writing:

> [The United States] allowed [itself] to be bluffed by China throughout most of the war. . . . Our error was not that we were fearful of the dangers of nuclear war and of Chinese or Russian intervention in Vietnam. These were proper concerns of the military strategist. The error was that we took counsel of these fears and in so doing paralyzed our strategic thinking.[7]

One of the other major obstacles to successfully fighting the war was a lack of agreement between the President (and his civilian advisors) and the military at the national level. Throughout the Johnson Administration each decision on the war seemed to be a compromise between the President's desire to maintain his domestic programs while defending South Vietnam and the Joint Chiefs of Staff's (JCS) belief that the only way to obtain a favorable solution was to terminate the restraints placed on military operations and strive for military victory. Summers was probably one of the harshest critics of the President's handling of the war effort in the decision-making process. He wrote:

> [A]s the Constitution envisions, the civilian leadership, the President and the Congress, must make the basic decisions about going to war and define the war's objectives. For their part, the nation's senior military leaders have the obligation to devise the strategy necessary for success, as they did in World War II and Korea. During the Indochina conflict, the U.S. Joint Chiefs of Staff (JCS) did not play this role. Unlike all his wartime predecessors, the President allowed civilian strategists with little or no combat experience to

---

[6] Dave Richard Palmer, *Summons of the Trumpet: U.S.-Vietnam in Perspective* (San Rafael, CA: Presidio Press, 1978), 110; quoted in Harry G. Summers, *On Strategy, A Critical Analysis of the Vietnam War* (New York: Dell Publishing Co., Inc., 1982), 127.

[7] Harry G. Summers, *On Strategy, A Critical Analysis of the Vietnam War* (New York: Dell Publishing Co., Inc., 1982), 94.

take charge, as if their "cost-benefit" or "psychological" approaches were superior to the insights of the military commanders.[8]

While Summers's criticism may be blunt, it maintains a thread of truth. Schandler, among others, confirmed the differing views between the President and his military advisors: "A debate behind closed doors concerning the limited strategy advocated by the President and his civilian advisors and the more forceful strategy advocated by the military chiefs continued intermittently throughout Lyndon Johnson's Presidency."[9]

Civilian strategy notwithstanding, the military leadership at the national level was certainly not without fault in directing operations during the war. Indeed, one of the most basic and logical principles in the conduct of war is to establish unity of command, which, in turn, should lead to a unity of effort. Such was not the case in the Vietnam theater of operations. In fact, command and control (C^2) was significantly fragmented, impacting reach, results, responsibility, and routes of communication. For example, during the war:

- Washington took a "business as usual" approach, with no organized mechanism established to oversee events.

- President Johnson became too involved in the day-to-day operations, even to the point of approving individual targets for air interdiction.

- The Army, Air Force, Navy, Marine Corps, and South Vietnamese all had their own air force, each answering to different commanders. The air war in the south was run by a headquarters in South Vietnam and in the north by the Commander in Chief, Pacific Command (CINCPAC) in Hawaii. Meanwhile, B-52 missions were conducted by Thailand and Guam remained under the control of the Strategic Air Command (SAC). In addition, interservice rivalries made cooperation between services difficult at best.

- CINCPAC, whose Area of Responsibility (AOR) included Vietnam, had very little to say in the conduct of ground

---

[8] Harry G. Summers, "Lessons: A Soldier's View," in *Vietnam as History, Ten Years After the Paris Peace Accords*, ed. Peter Braestrup (Washington, DC: University Press of America, 1984), 112.
[9] Schandler, "America and Vietnam," 24.

operations in South Vietnam although his position in the chain of command designated him as the theater commander. (Contrast this with General Schwarzkopf's performance in the Gulf War as Commander in Chief, Central Command (CINCCENT).) In fact, General William Westmoreland's headquarters in Saigon, theoretically required to deal with Washington through CINCPAC, often went directly to Washington.

- General Westmoreland had only coordination authority with the Vietnamese forces -- no combined command was established.

- General Westmoreland attempted to control all ground operations in South Vietnam from his headquarters in Saigon.

- Lower-level commanders believed there was too much control of the tactical operations by higher commanders; many senior commanders gave tactical orders from helicopters high above the action.

The lack of defined objectives from the civilian leadership translated into a lack of direction for military operations. Schandler opined:

> Paradoxically, General Westmoreland had adopted a basic concept of operations for which his civilian superiors in Washington would not provide the necessary operational leeway or, finally, the necessary level of forces. Westmoreland always hoped, as he stated in his memoirs, that the President would eventually allow him the authority to pursue the strategy, notably cutting the Ho Chi Minh Trail, that he and the Joint Chiefs saw as essential to a decisive outcome.[10]

Summers echoed Schandler's position when he wrote:

> The Joint Chiefs, led by General Earle Wheeler, strongly questioned the White House's approach [of a lack of strategy] in private, but Johnson (and Nixon) rarely consulted them directly. The Chiefs acquiesced in presidential mismanagement of the war, even allowing Johnson to set weekly bombing targets in North Vietnam; they hoped for better days.[11]

Ill-defined objectives from Washington translated into ill-defined military objectives in South Vietnam. Douglas Kinnard surveyed 173

---

[10] Ibid., 29.
[11] Summers, "Lessons: A Soldier's View," 112.

Army generals who had held command positions during the war and asked them the following question concerning U.S. objectives in Vietnam. Sixty-seven percent responded with the following results:

Were U.S. objectives in Vietnam prior to Vietnamization (1969):[12]

	Percent in Agreement
(1) Clear and understandable?	29
(2) Not as clear as they might have been?	33
(3) Rather fuzzy -- needed rethinking as the war progressed?	35
(4) Other?	3

It is a remarkable revelation that fully two-thirds of the United States military leadership during the Vietnam War were not totally clear on what the objectives in the war were.

This brief synopsis of selected strategic and operational problems encountered during the Vietnam conflict is not meant to rekindle debate over issues concerning that war. The intent is to show that in war, or in any given crisis, the basics on how the crisis should be dealt with can easily be lost in the "fog of war," even when those basics on how to deal with the situation are well known and understood. The tragedy of Vietnam, where the lack of strategic vision to guide the operational aspects of the war was lost on the leadership, can best be summed up as follows:

> The paradox arose of the Americans fighting on behalf of an army (and a government) that they treated with disdain, even contempt. The South Vietnamese were dealt with as if they were irrelevant. Thus there grew a naked contradiction between the political objectives of the war—an independent self-sufficient Vietnam—and the U.S. neglect of the South Vietnamese government and army in the formulation of American war strategy (or lack of it) during the 1965-67 build-up phases. There was little realization among American Unit leaders that crucial to South Vietnam's security was the development of an honest and efficient South Vietnamese government able and willing to improve the welfare of its people.

A final point needs to be made about Vietnam and the eventual conduct of future events. The young soldiers, sailors, and airmen who slogged their way through the rice paddies of Vietnam to make contact

---

[12] Douglas Kinnard, *The War Managers* (Hanover, NH: University Press of New England, 1977: reprint, New York: Da Capo Press, 1991), 24 (page reference is from reprint edition).

with the enemy, who were restricted from bringing maximum firepower to bear due to political restrictions, and who took the brunt of ill-defined directives in fighting the war were the same men who later formed the leadership cadre that directed the strategic actions, developed the campaign plans, and commanded the units involved in the Persian Gulf.

Chapter 2

Making Decisions

> To give meaning to the factual raw material of foreign policy, we must approach political reality with a kind of rational outline, a map that suggests to us the possible meanings of foreign policy.[13]

Classic Theory in Decision Making

The decision-making process for policy analysis at the seam of government between civilian leadership and its military counterpart is difficult at best. One has only to observe individuals from each culture at work within an American embassy in a foreign land to realize that each approaches a problem from an entirely different perspective, points of view that are inherent in their own personalities as well as *cultural* evolution in their respective professions. Fortunately, there is a thread of commonality in theory that helps bridge the gap between the divergent points of reference of the two groups. The dominant theory emphasizes a "rational" approach to the decision-making process, which falls directly in phase with the military concept of "estimate of the situation."

Public Administration

Decision-making theory as a field of research and study has been a prime contributor to public administration literature since the emergence of classical organizational theory itself; while the rational view of decision making traces its roots to the writings of Max Weber and Frederick Winslow Taylor.

---

[13] Hans J. Morgenthau, *Politics Among Nations*, 6th ed. (New York: Alfred A. Knopf, Inc., 1985), 5.

In his development of bureaucracy as an "ideal type," Weber defined the organization as "a system of authority relationships defined by rationally developed rules."[14] Weber characterized a bureaucracy as a rational and effective organization with the following characteristics:[15]

- Tasks are organized using rules.
- A division of labor produces specialization.
- Superior-subordinate relations are based on a hierarchy.
- Decisions are made by technical and legal standards.
- Administrative office holding is a vocation.

Weber later addressed one of the main points of politico-military decision making in his discussion on rationality. Using the term "instrumentally rational," he advanced the theory that an individual weighs the *means*, *ends*, and *consequences* of his actions as he makes decisions:

> A person acts rationally in the "means-end" sense when his action is guided by consideration of ends, means, and secondary consequences; when, in acting, he rationally *assesses* means in relation to ends, ends in relation to secondary consequences, and, finally the various possible ends in relation to each other.[16]

While Taylor did not write about rational decision making directly, his development of the scientific management concept instituted a rational and efficient approach to organizational operations and procedures. This idea of using specific scientific principles to order work correlates to the concept of decision making using the scientific method that evolved in the 19th century. In short, under both methods, problems can be solved by:

- Defining the problem.
- Collecting data.
- Developing alternative solutions.
- Analyzing the solutions.
- Selecting an alternative.

---

[14] Ralph C. Chandler and Jack C. Plano, *The Public Administration Dictionary* (New York: Macmillan Publishing Company, 1986), 154.
[15] Ibid., 155.
[16] Max Weber, *Max Weber, Selections in Translation*, ed. W.G. Runciman (Cambridge: Cambridge University Press, 1978), 29.

The most significant contributions to rational decision-making theory, however, evolved during the World War II era. During that time, Chester Barnard wrote *The Functions of the Executive* in which he described three roles of the executive. The third role applied directly to the decision-making process, with the executive "formulat[ing] and defin[ing] the purposes, objectives, ends of the organization."[17] It is these purposes and objectives that give direction to decision making and that, according to *Public Administration Dictionary* authors Ralph Chandler and Jack Plano, positioned Barnard as "the first major administrative theorist to develop the concept of the decision-making process. . . . [which] in Barnard's view, involves searching for strategic factors that meet the organization's purposes."[18]

A devotee of Barnard's work and one of the true classic theorists in public administration, Herbert Simon later addressed two very important aspects of rational decision making that deal with informational constraints and consequently have a significant impact on decision making in the military: "bounded rationality" and "satisficing." Both concepts deal with the lack of information when confronted with the need to make a decision; a serious problem in a crisis situation, particularly when military commanders must consider the disposition and intention of enemy forces as part of the process.

In his "rational man" model, Simon described the decision maker as one who "makes 'optimal' choices in a highly specified environment."[19] He wrote that a major impediment to good decision making, where one determines alternatives and selects the best course of action, is that rationality is constrained in three ways:

- Rationality requires a complete knowledge and anticipation of the consequences that will follow on each choice. In fact, knowledge of consequences is always fragmentary.

- Since these consequences lie in the future, imagination must supply the lack of experienced feeling in attaching value to them. But values can be only imperfectly anticipated.

---

[17] Chester I. Barnard, *The Functions of the Executive* (Cambridge, MA: Harvard University Press, 1938), 231.
[18] Chandler and Plano, 150.
[19] James G. March and Herbert A. Simon, *Organizations* (New York: John Wiley & Sons, 1958), 137.

- Rationality requires a choice among all possible alternative behaviors. In actual behavior, only a very few of all these possible alternatives ever come to mind.[20]

Simon called this constraint on rational decision-making "bounded rationality." The result of this limitation on information obliges the decision maker to draw conclusions based on imperfect knowledge of a given situation. He must, therefore, "satisfice" in the decision-making process by looking for "satisfactory" solutions rather than attempting to optimize the decision.

In later literature, Chandler and Plano described the "rational-comprehensive"[21] organizational decision-making model, first defined by Charles Lindblom. A systems analysis approach that is patterned on the scientific method for problem solving, it recognizes five steps in the decision-making process: defining the problem, developing alternative solutions, placing values on the consequences of each solution, assessing the probability that each alternative will occur, and making a decision based on logical rules. The model "attempts to serve the ideal embodied in Max Weber's view of bureaucracy in which decisions are based on impersonal rules and techniques."[22] Also based on the scientific method, Stephen Robbins's "optimizing model" encourages six similar steps to "maximize an outcome" in the decision-making process.[23]

At some point, however, basic decision-making theory must be integrated into policy-making theory. One of the earliest and most extensive works in this arena was *Foreign Policy Decision-Making* in which Richard C. Snyder, H.W. Bruck, and Burton Sapin analyzed the decision-making approach to international politics and determinants to state action.

Snyder, Bruck, and Sapin maintained that foreign policy decision making could best be analyzed in an organizational decision-making context. They contended that characteristics of any given organization and decision-making system were relatively the same, so that comparative analysis of different designs would provide insights into foreign policy making. However they also identified certain properties with respect to the norm in formal organizations that were unique and

---

[20] Herbert A. Simon, *Administrative Behavior: A Study of Decision-Making in Administrative Organization*, 3rd ed. (New York: The Free Press, 1976), 81.
[21] Chandler and Plano, 117.
[22] Ibid., 118.
[23] Stephen P. Robbins, *Essentials of Organizational Behavior*, 2nd ed. (Englewood Cliffs, NJ: Prentice Hall, 1988), 59.

applied directly to policy formulation, thereby contributing to an understanding of how the foreign policy-making process functioned. In organizations concerned specifically with policy making, they found:

- A wider range of *possible objectives and projects* subject to a wider range of *possible interpretations*.

- A greater *heterogeneity of "clientele"* and thus more potentially hostile or dissatisfied *reactions* and *demands*.

- That a greater *number of perspectives* have to be *integrated* before consensus is achieved.

- That the "*setting*" and "*situation*" of decisions are more complex, less certain, less stable; therefore the consequences of action are *harder to predict and control*.

- That s*ources of information* are broader and less reliable, and the necessity of "classification" constitutes a special problem.

- A relative lack of "*experimental opportunity*" and infrequency of replicable situations.

- That it is difficult to measure *organizational effectiveness* and *policy results*.

- That alternatives often must be discussed in terms that *do not* meet the simplest test of *verifiability*.

- A *time lag* between the development of a problem-situation and the *unfolding of its full implications*.

- A greater possibility of *fundamental value conflicts* and hence necessity *for more extensive compromise*.[24]

In short, they uncovered the inherent degree of uncertainty involved in the foreign policy-making process.

Perhaps the most read and widely regarded effort at decision-making and policy analysis is Graham Allison's examination of the Cuban Missile Crisis and his explanation of the three conceptual models of

---

[24] Richard C. Snyder, H.W. Bruck, and Burton Sapin, eds., *Foreign Policy Decision-Making* (New York: The Free Press of Glencoe, 1962), 104.

government behavior in *Essence of Decision*. Allison called his first model the "Rational Actor Model," which he maintained is the most widely used decision-making process in the conduct of foreign policy. The model follows the scientific method of problem solving and can be characterized by asking the following questions:

- What is the problem?
- What are the alternatives?
- What are the strategic costs and benefits associated with each alternative?
- What is the observed pattern of national (governmental) values and shared axioms?
- What are the pressures in the "international strategic marketplace"?[25]

In the same terms as those embraced by many classical and neoclassical writers, the process involves setting objectives, developing alternatives, examining the consequences, and selecting the alternative that best achieves the objectives. "Rationality," Allison stated, "refers to consistent, value-maximizing choice within specified constraints."[26]

When researching decision-making theory, particularly in the policy arena where staffs, committees, task forces, or other groups are involved, the concept of "groupthink" identified by Irving Janis must be regarded as a significant impediment to formulating optimal decisions. Janus cited several characteristics of groupthink that precluded members of a group from clearly analyzing a problem. These include:

- An illusion of invulnerability that creates excessive optimism and encourages taking risks.
- A belief in the group's inherent morality, creating a propensity to ignore ethical or moral consequences.
- Efforts to discount warnings or other information that would lead members to reconsider their assumptions.
- Stereotyped views of the enemy as too evil to warrant attempts at negotiations.
- Self-censorship of those who may deviate from the group.
- A shared illusion of unanimity concerning decisions conforming to the majority.

---

[25] Graham T. Allison, *Essence of Decision* (Boston: Little, Brown and Company, 1971), 257.
[26] Ibid., 30.

- Self-appointed "mindguards" who protect the group from adverse information relative to the morality or effectiveness of their decisions.[27]

If a policy-making group began to exhibit most or all of these characteristics, Janus maintained that the group's effectiveness had the potential to degrade significantly. He considered the Bay of Pigs *faux pas* an example of a "perfect failure" based on the power of groupthink on the decision-making process. In what appears to be a testimony to the failure of Vietnam, he listed a number of problems associated with decision making that could be accentuated by the groupthink syndrome, including:

- Incomplete survey of alternatives.
- Incomplete survey of objectives.
- Failure to examine risks of preferred choice.
- Failure to reappraise initially rejected alternatives.
- Poor information search.
- Selective bias in processing information at hand.
- Failure to work out contingency plans.[28]

A Military Context

While the civilian arm of government may view a potential problem from a different perspective, the decision-making *process* the leadership uses is theoretically in step with the military, who also applies a rational approach to decision making.

The *Department of Defense Dictionary* describes the military's "estimate of the situation" as a "logical process of reasoning by which a commander considers all the circumstances affecting the military situation and arrives at a decision as to the course of action to be taken in order to accomplish his mission."[29] The process can be, and will be shown to be, useful for decision making at all three levels of war (strategic, operational, and tactical), only the scope and detail varies among the levels. For example, the estimate process focuses on "what to do" and "how to do" an operation in support of a crisis or problem. At the strategic level emphasis is placed more on the "what to do,"

---

[27] Irving L. Janis, *Groupthink*, 2nd ed. (Boston, MA: Houghton Mifflin Company, 1982), 174.
[28] Ibid., 175.
[29] *Department of Defense Dictionary*, s.v. "Estimate of the situation."

whereas at the tactical level more weight is placed on the "how to do it." The commander at the operational level must focus on both.

The estimate of the situation method is taught at military senior service schools such as the National War College (NWC) and the Industrial College of the Armed Forces (ICAF). The schools' curricula characterize the estimate approach in basically the same manner as public administration literature explains scientific problem solving and policy analysis. The steps in the estimate method are as follows:

- Analyze the mission.
- Estimate the friendly, enemy, and environmental situations.
- Correlate combat power and other influences.
- Devise courses of action (possibilities).
- Analyze courses of action against enemy courses and among themselves (consequences).
- Choose one.

There are two factors that are critical to the estimate of the situation: opposed versus unopposed time, and dynamic circumstances. When the military establishment is brought into a foreign policy issue, it is usually because a situation has materialized where *confrontation* with another nation or truculent actor is a real possibility. With real-time opposition as a factor in the decision process, a whole host of additional issues must be considered by the leadership, many involving life and death considerations. The opposite can be said for the time factor. Often when military options are being considered, both the civilian and military leadership may have a finite amount of time to react, often without enough information available to make an insightful decision. To the same extent, conditions inherent in a crisis are dynamic in nature, and the circumstances surrounding a given situation can change in magnitude or direction at any given moment. Once a decision has been made, it is essential that the decision maker reevaluate the estimate to ascertain what has changed. More often than not, the commander must reenter the decision-making process, often using it as an iterative process.

A number of classical military theorists through the ages have addressed the decision-making process based on their inquiry into strategy and doctrine. For example, Sun Tzu in 500 B.C. stressed several decision-making factors that should be considered when confronting an adversary. Machiavelli was very specific on the nature of decision making in *The Discourses* when he discussed the principle of unity of command. Military decision making will therefore be further defined within a strategic theory context when strategy and doctrine are brought together in the next chapter.

Principles in Public Administration

The scientific management school in classical public administration theory featured a rational approach to decision making. Inherent in this concept was the desire to simplify processes down to their basic elements. The basis for this simplification approach was the search for fundamental tenets, or "principles," to explain and/or guide the practical application of theory. Hence, early writers of public administration used the very same techniques that currently exist in military strategic theory to explain functions, relationships, and procedures in public administration theory. In fact, a parallel can be drawn between the two fields in the way their theoretical foundations evolved, particularly during their early, formative years.

Although not the first, Leonard White was one of the early public administration theorists to support the establishment of principles as essential to instituting order in organizational procedures. He wrote: "Contrary to the politician, whose principles might be tempered by the changing winds of opinion, the administrator cherishes the pattern of consistency. . . . He believes himself guided by general views, by settled tradition, by established knowledge of appropriate methods in administrative operations."[30] White defined principle as "a hypothesis so adequately tested by observation and/or experiment that it may intelligently be put forward as a guide to action, or as a means of understanding."[31] He then acknowledged the importance of employing principles in administration when he wrote:

> The history of government reveals a constant series of great figures who out of experience formulated hypotheses and came to personal convictions as to principles. Richelieu, Burke, Hamilton, John Stuart Mill, Charles Francis Adams, Lord Haldane, Graham Wallas, Henri Fayol, these are among the great names who in their day developed programs of public administration.[32]

---

[30] Leonard White, "The Meaning of Principles in Public Administration," in *The Frontiers of Public Administration*, ed. John M. Gaus, Leonard D. White, and Marshall E. Dimock (Chicago: The University of Chicago Press, 1936; reissue, New York: Russell & Russell, 1967), 13.
[31] Ibid., 21.
[32] Ibid., 21.

A contemporary of White, Marshall Dimock, also supported the need to employ principles in the field of administration. He asserted that "even if only a small number of time-tested verities can be relied upon, it must be insisted that an objective of public administration is to follow such principles as consistently as possible and constantly to be on the lookout for the testing and enunciation of new ones."[33]

The British scientist and mathematician Charles Babbage was one of the first to conclude that there were basic principles of management that could be broadly applied in the field. In 1832 he defined these principles in his book *On the Economy of Machinery and Manufactures*, which influenced Frederick Taylor and the scientific management movement 60 years later.[34] Taking his cue from Adam Smith, Babbage described his principle of "division of work" as follows:

> That the master manufacturer, by dividing the work to be executed into different processes, each requiring different degrees of skill or of force, can purchase exactly that precise quantity of both which is necessary for each process; whereas, if the whole work were executed by one workman, that person must possess sufficient skill to perform the most difficult, and sufficient strength to execute the most laborious, of the operations into which the art is divided.[35]

Henri Fayol, a French industrialist, was the first to develop a comprehensive theory of management, which he believed to be suitable for use in all organizations. In his book *General and Industrial Management* he gave great credence to the relevance of principles and the importance of using them in what he labeled as the management "code":

> Principles are flexible and capable of adaptation to every need; it is a matter of knowing how to make use of them, which is a difficult art requiring intelligence, experience, decision and proportion . . . There

---

[33] Marshal E. Dimock, "The Criteria and Objectives of Public Administration," in *The Frontiers of Public Administration*, ed. John M. Gaus, Leonard D. White, and Marshall E. Dimock (Chicago: The University of Chicago Press, 1936; reissue, New York: Russell & Russell, 1967), 129.

[34] Jay M. Shafritz and J. Steven Ott, *Classics of Organization Theory*, 2nd ed. (Chicago: The Dorsey Press, 1987), 23.

[35] Charles Babbage, *On the Economy of Machinery and Manufactures* (London: Charles Knight, Pall Mall, East, 1835; reprint, New York: Augustus M. Kelly, Bookseller, 1963), 175-176 (page references are to the reprint edition).

is no limit to the number of principles of management, every rule or managerial procedure which strengthens the body corporate or facilitates is functioning has a place among the principles so long, at least, as experience confirms its worthiness.

This code is indispensable. Be it a case of commerce, industry, politics, religion, war, or philanthropy, in every concern there is a management function to be performed, and for its performance there must be principles, that is to say acknowledged truths regarded as proven on which to rely. And it is the code which represents the sum total of these truths at any given moment.[36]

The 14 principles of management that Fayol identified as being most useful to him were:[37]

1. Division of labor
2. Authority
3. Discipline
4. Unity of command
5. Unity of management
6. Subordination of individual interests to the common good
7. Remuneration
8. Centralization
9. The hierarchy
10. Order
11. Equity
12. Stability of staff
13. Initiative
14. Esprit de corps

Frederick Taylor, in his time-and-motion studies leading to the adoption of scientific management as a movement around 1910, cited several principles that would lead to the "one best way" for the efficient production of goods in the factory. He proclaimed these as the "great principles of scientific management," and identified them as follows:

- Develop a science for each element of work that replaces a rule-of-thumb method.
- Scientifically select and then train, teach, and develop the workmen.
- Cooperate with the workmen to insure work is done according to the principles of science.
- Equally divide the work and responsibility between management and the workmen.[38]

---

[36] Ibid., 19, 41-42.
[37] Ibid., 19.
[38] Frederick Winslow Taylor, *The Principles of Scientific Management* (Norwood, MA: The Plimpton Press, 1911; reprint, New York: Harper & Brothers Publishers, 1934), 36-37 (page references are to reprint edition).

Max Weber, in his inquiry into the characteristics of bureaucracies, identified principles he claimed were inherent in organizational design. First of all, in his "ideal type" of core features in a bureaucracy, he defined three "pure types" of legitimate authority: legal, traditional, and charismatic. Second, in describing the central characteristics of the bureaucratic institution (admittedly, somewhat stretching the "principles" point, but it is modeled on the concept of facilitating basic awareness and understanding), he related how administrators operate according to 14 principles and criteria intrinsic to a hierarchical organization.

James Mooney agreed with Fayol in the universal applicability of utilizing principles in organizations. In his discussions on industrial organization, Mooney stated that "the term organization, and the principles that govern it, are inherent in every form of concerted human effort, even where there are no more than two people involved."[39]

Although not specifically identifying them as principles, Luther Gulick, in a concept he adapted from Fayol, illustrated the primary functions of the chief executive with his acronym POSDCORB: *planning, organizing, staffing, directing, coordinating, reporting,* and *budgeting*. Gulick professed his belief in the universality of his theory when he wrote that "it is believed that those who know administration intimately will find in this analysis a valid and helpful pattern, into which can be fitted each of the major activities and duties of any chief executive."[40] One of the major consequences resulting from Gulick's work was the emphasis placed on "division of work" as the foremost principle for attaining organizational production goals.

Following the end of the Second World War Herbert Simon identified four of the more common management principles cited in public administration literature:

- Administrative efficiency is increased by a specialization of the task among the group.

- Administrative efficiency is increased by arranging the members of the group in a determinate hierarchy of authority.

---

[39] James D. Mooney, "The Principles of Organization," in *Papers on the Science of Administration*, ed. Luther Gulick and L. Urwick (New York: Institute of Public Administration, 1937), 91.

[40] Luther Gulick, "Notes on the Theory of Organization," in *Papers on the Science of Administration*, ed. Luther Gulick and L. Urwick (New York: Institute of Public Administration, 1937), 13.

- Administrative efficiency is increased by limiting the span of control at any point in the hierarchy to a small number.

- Administrative efficiency is increased by grouping the workers, for purposes of control, according to (a) purpose, (b) process, (c) clientele, or (d) place.[41]

Rather than extolling the virtues of employing these principles in the administrative process, however, he took a more critical view of their use, writing in *Administrative Behavior*:

> It is a fatal defect of the current principles of administration that, like proverbs, they occur in pairs. For almost every principle one can find an equally plausible and acceptable contradictory principle. Although the principles of the pair will lead to exactly opposite organizational recommendations, there is nothing in the theory to indicate which is the proper one to apply.[42]

Simon asserted that principles identified in earlier literature were not explicit enough because they lacked definition and, although they may have provided choices, they were not clear on which option should be chosen.

Simon did not reject the notion of using principles out of hand; he was just critical of the way earlier principles had been applied in public administration theory. Rather than focusing on terminology that he considered to be too ambiguous, he believed that the study of principles:

> Requires that *all* the relevant diagnostic criteria be identified; that each administrative situation be analyzed in terms of the entire set of criteria; and that research be instituted to determine how weights can be assigned to the several criteria when they are, as they usually will be, mutually incompatible.[43]

As can readily be seen, Simon's position on principles in administration bridges the gap between the singular concept of applying principles in decision making and the much more complex theories involving decision making itself. In fact, Simon believed that a manager or administrator should not accept the application of principles

---

[41] Simon, 20-21.
[42] Ibid., 20.
[43] Ibid., 36.

as an end in itself, but as part of the overall process for determining the correct decision in a given situation.

A Systems Approach?

A systems approach to management (and its various adaptations) remains one of the most widely accepted concepts in the leadership and management fields today. By modifying the theory slightly, the systems approach in general and contingency theory in particular set the stage for merging public administration, military decision making and the use of principles together into a concept for practical application.

The systems approach was first characterized by Ludwig von Bertalanffy, later given wide acceptance through the writings of Daniel Katz and Robert Kahn, and then modified to include uncertainty by contingency theorists. (See Figure 2.1.)

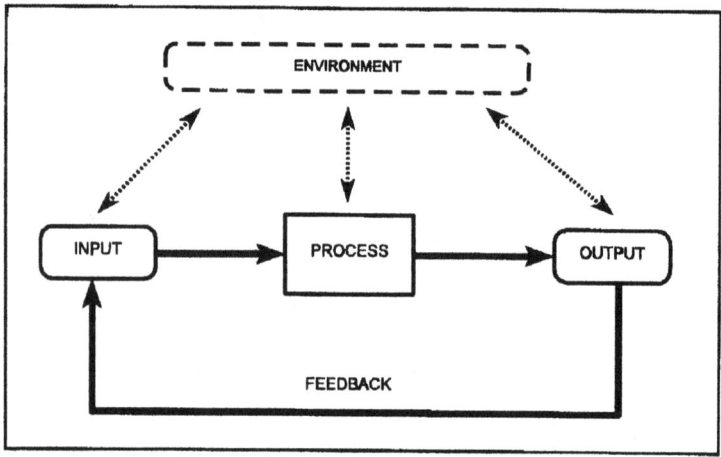

Figure 2.1 The Systems Approach

Von Bertalanffy gave the approach a sense of "wholeness" by integrating a set of complex elements, including inputs, processes, and outputs that form the basis of organizational operations. Katz and Kahn emphasized the need to open the system to the influence of its environment. James Thompson then provided the foundation for contingency theory in an open system when he wrote:

If the organization must approach certainty at the technical level to satisfy its rationality criteria but must remain flexible and adaptive to satisfy environmental requirements, we might expect the managerial level to mediate between them, ironing out some irregularities stemming from external sources but also pressing the technical core for modifications as conditions alter.[32]

The systems approach can be used to illustrate the decision-making process as portrayed in Figure 2.2. In addition, Thompson's organization as an open system provides an appropriate vehicle to identify where principles would be applied in the process between the civilian and military leadership at the national level of government. Thompson described his concept of open system organizations as follows: *"We will conceive of complex organizations as open systems, hence indeterminate and faced with uncertainty, but at the same time as subject to criteria of rationality and hence needing determinateness and certainty."*[33] By incorporating the use of principles as part of the decision-making process as seen in Figure 2.2, the ability to reduce the effect of uncertainty is enhanced.

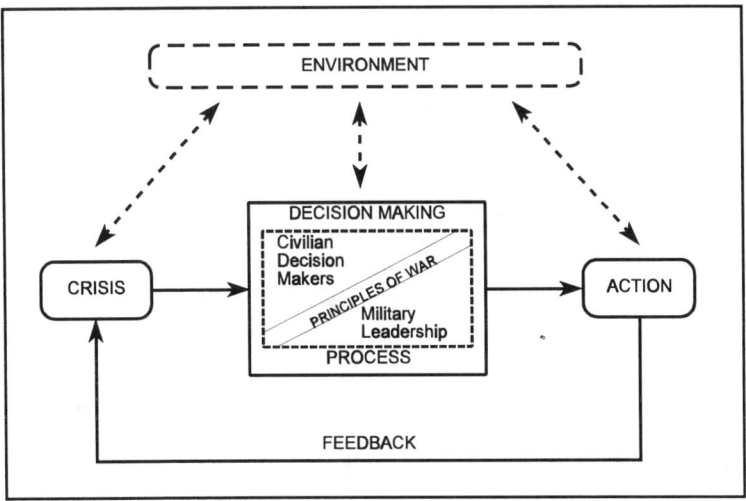

Figure 2.2 Crisis Decision Making

---

[32] James D. Thompson, *Organizations in Action* (New York: McGraw-Hill Book Company, 1967), 12.
[33] Ibid., 10.

Principles cannot, however, be used as inflexible checklists or rules that are expected to work, without fail, in every case without adaptation to the situation. They can, though, be used as a *guide* to plan the response to a given situation; there is no one best way to address all contingencies. Each situation deserves to be dealt with based on the particular conditions at that particular time.

The military approach to decision making can be illustrated by employing decision-making theory in a practical situation (Figure 2.3). When faced with a threat and given objectives by civilian leadership to counter a crisis, a military commander makes an estimate of the situation and has his staff begin initial preparations.[34] As the process of developing a plan and mobilizing forces advances through its first cycle, the commander and his staff continue to assess the developing situation. Once the plan has been executed, any change in the situation is identified and appropriate action is taken in the next cycle to counter the resulting outcome. Taught to senior military officers and civilians at National Defense University, this process mirrors the systems approach model illustrated in Figure 2.1.

---

[34] The authors fully appreciate the fact that in today's military command billets and other traditionally male-dominated positions can be occupied by either sex. Reference to the masculine gender with respect to those positions is for succinctness only and is not meant to detract from the important role females play in the military today.

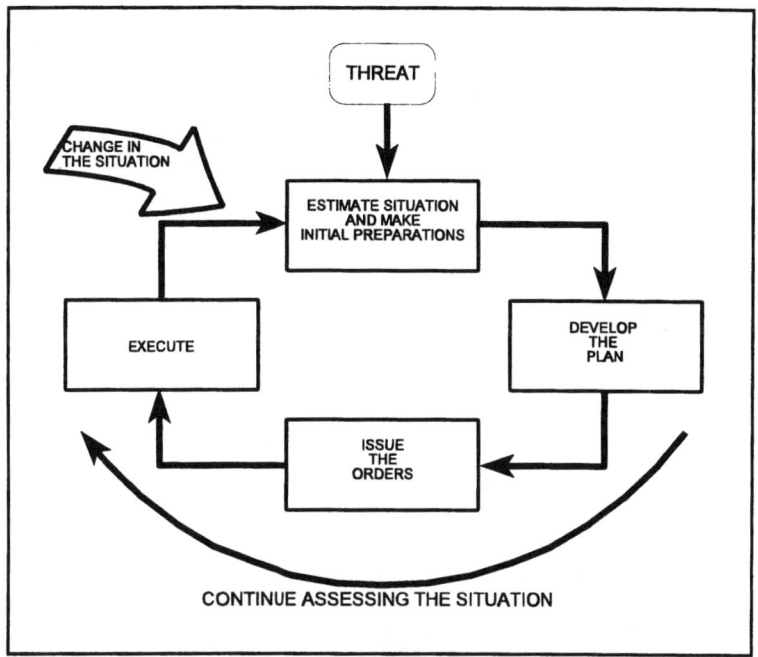

Figure 2.3 Action Cycle

Chapter 3

The U.S. Military and the Art of War

In the study of military strategic thought one must examine works by the great military writers. The use of these theoreticians is extremely important today because a great deal of what they have written remains relevant despite extensive changes in warfare. In fact, as the post-World War II military institution in the United States discounted the value of several of these seminal works, the national security establishment lost sight of how to properly prosecute war, and paid a terrible price in the end. Only in the past 10 to 15 years have these theoretical roots been revisited, resulting in big dividends in the Gulf War.

Military Theorists

In the study of strategy, particularly military strategy, the classical writers have withstood the test of time. The works of Sun Tzu, Machiavelli, Jomini, and particularly Carl von Clausewitz are used to this day as points of departure for discussing the art of war. Certainly later theorists have also contributed to the discussion, but they have not strayed far from classical thought. Prominent during the latter part of the 19th century and the post-World War I period, writers such as Mahan, Fuller, Hart, and Douhet, for example, developed their own concept of strategic theory that was mostly supportive of, sometimes critical of, but usually only attempted to incrementally change the basics of classical thought. Modern writers, including Brodie, Eccles, and Wylie, have also only added a more contemporary slant to the body of knowledge based on changes in prevailing doctrine, rather than trying to discredit the past. The classical theorists have indeed withstood the test of time.

To establish how the principles of war relate to strategic theory, first strategy, then doctrine (a subset of strategy), and finally the principles of war themselves (a subset of doctrine) will be examined from a classical and contemporary theoretical context, grounded primarily on the writings of the strategic thinkers listed above. This strategic theory will then be placed in an operational framework predicated on the "levels of war." "War termination" will also be addressed since it should influence the application of strategy.

Strategic Theory

Sun Tzu wrote his classic text, *The Art of War*, over 2,500 years ago, yet it remains one of the widest read and highest regarded works on strategic theory. Liddell Hart described Sun Tzu's influence as follows:

> Sun Tzu's essays on 'The Art of War' form the earliest of known treatises on the subject, but have never been surpassed in comprehensiveness and depth of understanding. They might well be termed the concentrated essence of wisdom on the conduct of war. Among all the military thinkers of the past, only Clausewitz is comparable, and even he is more 'dated' than Sun Tzu, and in part antiquated, although he was writing more than two thousand years later. Sun Tzu has clearer vision, more profound insight, and eternal freshness.[35]

Sun Tzu's predilection toward war was one of extreme prudence: "War is a matter of vital importance to the State; the province of life or death; the road to survival or ruin. It is mandatory that it be thoroughly studied."[36] He did not believe that the ultimate objective of war was the annihilation of the enemy in battle. To him, all warfare should be based on strategy and deception; he therefore placed a high value on the importance of strategic thought. "Thus, those skilled in war subdue the enemy's army without battle. They capture his cities without assaulting them and overthrow his state without protracted operations."[37] In fact, Sun Tzu believed that strategy was one element that spelled the difference between victory and defeat. "A victorious army wins its victories before seeking battle; an army destined to defeat fights in the hope of winning,"[38] he wrote. Liddell Hart agreed with

---

[35] Sun Tzu, *The Art of War*, trans. Samuel B. Griffith (New York: Oxford University Press, 1971), v.
[36] Ibid., 63.
[37] Ibid., 79.
[38] Ibid., 87.

Sun Tzu on the aim of strategy. "The perfection of strategy would be, therefore, to produce a decision without any serious fighting."[39]

Sun Tzu's junior by 2,000 years, Niccolò Machiavelli (strategy was a major theme in his writings on the art of war) maintained that the aim of war is to subject the enemy to your will; therefore, he argued, a military campaign must be planned.[40] According to Machiavelli, the correct strategy for carrying out this aim depends on the particular circumstances under which the campaign is conducted.

Baron Henri Jomini took this argument one step further, explaining strategy in relation to logistics and tactics:

> Strategy is the art of making war upon the map, and comprehends the whole theater of operations. Grand tactics is the art of posting troops upon the battlefield according to the accidents of the ground, of bringing them into action. . . . Logistics comprises the means and arrangements which work out the plans of strategy and tactics. Strategy decides where to act; logistics brings the troops to this point; grand tactics decides the manner of the execution and the employment of the troops.[41]

Liddell Hart offered Clausewitz's definition of strategy as "the art of the employment of battles as a means to gain the object of war. In other words, strategy forms the plan of the war, maps out the proposed course of the different campaigns which compose the war, and regulates the battles to be fought in each."[42]

Hart was critical, and rightly so, of this definition because it constrained the art of strategy by associating it too closely with the actual conduct of the war at the tactical level. In reality the concept needs to be broadened to include the strategic level of war, and to a lesser extent, the operational level. Hart accomplished this to some degree by focusing the definition on the civilian/military seam of government by defining strategy as "the art of distributing and applying military means to fulfill the ends of policy."[43] He then went

---

[39] B.H. Liddell Hart, *Strategy*, 2nd revised ed. (New York: NAL Penguin Inc, 1974), 324.

[40] Felix Gilbert, "Machiavelli: The Renaissance of the Art of War," in *Makers of Modern Strategy From Machiavelli to the Nuclear Age*, ed. Peter Paret (Princeton: Princeton University Press, 1986), 29.

[41] Baron Henri Jomini, *The Art of War*, trans. by Capt. G.H. Mendell and Lieut. W.P. Craighill (Philadelphia, PA: J.B. Lippincott & Co., 1862; reprint, Westport, CT: Greenwood Press, 1971), 69 (page references are to reprint edition).

[42] Hart, 319.

[43] Ibid., 321.

on to broaden the concept further by addressing "grand strategy," which, by his definition, takes into account all instruments of national power required to prosecute national policy while looking at the peace following war termination. In the end, Hart agreed with Sun Tzu that the perfection of strategy would be to gain a decision in a crisis without any serious fighting.[44]

There is no doubt that because of his army background Clausewitz focused on the tactical employment of strategic thought, particularly with his emphasis on continental warfare. This should not diminish his standing as a strategic thinker, however, nor degrade the usefulness of his theories at the strategic and operational levels of war. Clearly then, with respect to Hart's criticism, Clausewitz did broaden his concept of strategy, noting:

> The strategist must, therefore, define an aim for the entire operational side of the war that will be in accordance with its purpose. In other words, he will draft the plan of the war, and the aim will determine the series of actions intended to achieve it . . . Strategic theory, therefore, deals with planning.[45]

Clausewitz expanded his concept of strategy even further as he focused on the relationship linking the government, armed forces, and citizens. "A theory that ignores any one of them or seeks to fix an arbitrary relationship between them would conflict with reality to such an extent that for this reason alone it would be totally useless,"[46] he argued.

Clausewitz related strategy to national policy and saw war as an instrument of that policy. Based on his supposition that "war . . . is a continuation of political intercourse, carried on with other means,"[47] it follows that policy objectives form the basis of strategic formulation. Without policy there can be no strategy, and without strategy war cannot exist as a continuation of policy.

Clausewitz then addressed the operational aspects of war when he defined strategy as "the use of engagements for the object of the war."[48] Just as he identified the political object as the aim of strategy at the national level, he identified the object of strategy in a theater context as the enemy's center of gravity. "One must keep the dominant characteristics of both belligerents in mind. Out of these characteristics

---

[44] Ibid., 324.
[45] Clausewitz, 177.
[46] Ibid., 89.
[47] Ibid., 87.
[48] Ibid., 128.

a certain center of gravity develops, the hub of all power and movement, on which everything depends. That is the point against which all our energies should be directed."[49]

French military theorist André Beaufre did not distinguish between the concepts of "strategy" and "grand strategy" as defined by Hart (which is more consistent with the dominant view today). Believing that Hart was too restrictive in his definition, Beaufre defined strategy as "the art of applying force so that it makes the most effective contribution towards achieving the ends set by political policy."[50] He went on to clarify his position as he explained that the aim of strategy "is to fulfill the objectives laid down by policy, making the best use of the resources available."[51] In his reference to "resources," Beaufre was referring to all instruments of national power.

Bernard Brodie kept his fundamental concept of strategy fairly simple and succinct: "Strategy is a 'how to do it' study, a guide to accomplishing something and doing it efficiently. As in many other branches of politics, the question that matters in strategy is: Will the idea work? . . . Above all, strategic theory is a theory for action."[52]

Henry Eccles took more or less the same stance as previous writers, which is indicative of a trend toward an agreed-upon position. According to Eccles:

> Strategy is the comprehensive direction of power to control situations and areas to attain broad objectives . . . Strategy is comprehensive, it aims sat the whole field of action. Strategy is the direction of power. It should not concern itself with operational details. Strategy deals with all forms of power available to the command.[53]

Eccles supported the classical works when he added, "The first principle of strategy is that political purpose must dominate strategy. The use of military force without a clear political purpose is ultimately futile and self-defeating."[54]

---

[49] Ibid., 595.

[50] André Beaufre, *An Introduction to Strategy* (London: Faber and Faber, 1965), 22.

[51] Ibid., 23.

[52] Bernard Brodie, *War & Politics* (New York: Macmillan Publishing Co., Inc., 1973), 452.

[53] Henry E. Eccles, "Strategy, The Theory and Application," in *Military Strategy: Theory and Application*, ed. Colonel Arthur F. Lykke, Jr. (Ret.), (Carlisle Barracks, PA: U.S. Army War College, 1989), 37.

[54] Ibid., 37.

Another modern writer, Major General E.B. Atkeson, contributed to strategic theory in his well-regarded book *The Final Argument of Kings*, identifying five basic approaches to strategy for analytical interest. He described them as:

- The classical (or historical) approach provides the basic language and relationships of strategic thought; identifies strategic principles, or axioms; and introduces a well-developed field of honored philosophers.

- The spatial approach deals with geographic questions on the battlefield, as well as global proportions, both on the sea and in the air.

- The power potential approach compares military forces and mobilization power of potential adversaries, perhaps the most widely used approach to strategic analysis today. This approach can go beyond military strength to include the true power of a nation associated with its political, economic, and psychosocial fabric.

- The technological approach not only deals with the strategic application of technology, but also considers the adaptation of strategy, organization, and doctrine to accommodate technological change.

- The ideological/cultural approach relates to the ideological and cultural values of a society and its penchant for identifying with states having similar political or ethical dispositions.[55]

For the purposes of this analysis, the definition of strategy found in the *Department of Defense Dictionary of Military and Associated Terms* will be used as a point of departure. Strategy is:

The art and science of developing and using political, economic, psychological [now often referred to as informational], and military forces as necessary during peace and war, to afford the maximum support to policies, in order to increase the probabilities and

---

[55] E.B. Atkeson, Maj. Gen., U.S. Army (Ret.), *The Final Argument of Kings, Reflections on the Art of War* (Fairfax, VA: Hero Books, 1988), 65-74.

favorable consequences of victory and to lessen the chances of defeat.[56]

The concept of strategy can be broadened by considering two additional terms, national strategy and military strategy. National strategy is defined as: "the art and science of developing and using the political, economic, and psychological powers of a nation, together with its armed forces, during peace and war, to secure national objectives."[57] Military strategy is defined as: "the art and science of employing the armed forces of a nation to secure the objectives of national policy by the application of force or the threat of force."[58] When considering military strategy one must ensure that the ends (or objectives) and the ways and means are consistent with the political objectives. Furthermore, they must be achievable using the resources available to the nation and must be designed to focus on a potential enemy's center of gravity.

Military Doctrine

According to the *Department of Defense Dictionary*, doctrine is "fundamental principles by which the military forces or elements thereof guide their actions in support of national objectives. It is authoritative but requires judgment in application."[59] *Joint Pub 1, Joint Warfare of the US Armed Forces* provides a more comprehensive perspective on the role of doctrine in the U.S. military based on the philosophy of the current leadership:

Military doctrine presents fundamental principles that guide the employment of forces. **Doctrine is authoritative but not directive** [emphasis added]. It provides the distilled insights and wisdom gained from our collective experience with warfare. However, **doctrine cannot replace clear thinking or alter a commander's obligation to determine the proper course of action** [emphasis added] under the circumstances prevailing at the time of decision.[60]

---

[56] *Department of Defense Dictionary*, s.v. "Strategy."
[57] Ibid., s.v. "National strategy."
[58] Ibid., "Military strategy."
[59] Ibid., s.v. "Doctrine."
[60] Department of Defense, *Joint Pub 1, Joint Warfare of the US Armed Forces* (Washington, DC: Department of Defense, The Joint Chiefs of Staff, 11 November 1991), 5.

The Joint Pub 1 explanation reveals two features inherent in the United States' concept of doctrine. First, doctrine is not made up of hard and fast rules; the skill and experience of a leader to make the correct choice when faced with a critical decision make up a fundamental tenet. Second, lessons learned from the past based on experience and through the writings of others should not be discounted, even with the monumental changes that have taken place in the conduct of war during the 20th century.

Clausewitz championed this need for knowledge and experience as a requirement for success in leadership positions:

> The knowledge needed by a senior commander is distinguished by the fact that it can only be attained by a special talent, through the medium of reflection, study and thought: an intellectual instinct which extracts the essence from the phenomena of life, as a bee sucks honey from a flower. In addition to study and reflection, life itself serves as a source. Experience, with its wealth of lessons, will never produce a *Newton* or an *Euler*, but it may well bring forth the higher calculations of a *Condé* or a *Frederick*.[61]

Although Clausewitz had trouble coming to grips with "hard and fast rules" in the development of strategy and doctrine, he recognized the value of formulating principles when used in the manner described in Joint Pub 1. Using historical examples to build a knowledge base in the art of war, Clausewitz related their use in the grounding of doctrine: "The detailed presentation of a historical event, and the combination of several events, make it possible to deduce a doctrine: the proof is in the evidence itself."[62]

Debate continues over whether or not the applicability of doctrine should follow hard and fast rules. In the end, however, when one examines the intent of most classical and contemporary writers, they tend to agree that doctrinal principles cannot be "cast in stone."

Often criticized for his attempts to establish scientific, non-varying rules in the conduct of war, Jomini realized that strict adherence to such tenets should not be the case. In concluding his thesis on the art of war he stated:

> It is true that theories cannot teach men with mathematical precision what they should do in every possible case; but it is also certain that they will always point out the errors which should be avoided; and this is a highly-important consideration, for these rules thus become,

---

[61] Clausewitz, 146.
[62] Ibid., 171.

in the hands of skillful generals commanding brave troops, means of almost certain success.[63]

Contemporary writers have also addressed the merits of doctrine. General Curtis E. LeMay, father of Strategic Air Command, gave an excellent description on the role and value of doctrine as follows: "At the very heart of war lies doctrine. It represents the central beliefs for waging war in order to achieve victory. . . . It is the building material for strategy. It is fundamental to sound judgment."[64]

Well-respected contemporary writers in the field of strategic theory Dennis Drew and Donald Snow reduced the concept of doctrine down to its very essence: "Military doctrine is what we believe about the best way to conduct military affairs."[65] They go on to write that the principal source of doctrine is experience. Their definition coincides with current accepted thinking, which helps to legitimate positions taken by both classical and modern writers over the years and provides a solid foundation for examining the principles of war.

Drew and Snow also presented an important view on the relationship between doctrine and strategy. They stated that doctrine embodies four functions that aid leaders in making strategic decisions. Namely, it:

- Provides a tempered analysis of experience and a determination of beliefs.
- Teaches those beliefs to each succeeding generation.
- Provides a common basis of knowledge and understanding that can provide guidance for actions.
- Provides a standard against which to measure efforts, an indicator of success and a tool for analyzing both success and failure.[66]

The first three functions have been addressed by other theorists and are accepted by the field. The fourth function is unique and provides the basis for examining the principles of war as they relate to the wars in Angola and the Gulf.

If the success or failure of a strategy is determined to rest on the doctrine employed, then that doctrine will endure in the case of success

---

[63] Jomini, 323.

[64] Curtis E. LeMay, General, Air Force Manual 1-1, *Basic Doctrine* (Washington, DC: Department of the Air Force, 1984), frontispiece, quoted in Joint Pub 1, 5.

[65] Dennis M. Drew and Donald M. Snow, *Making Strategy; An Introduction to National Security Processes and Problems* (Maxwell Air Force Base, AL: Air University Press, 1988), 163.

[66] Ibid., 171-173.

but will require modification in the case of failure. It is this fourth function that establishes the relationship of doctrine to strategy, and Drew and Snow illustrated this bond when they wrote, "Doctrine influences strategy (or it should) and the results of strategy become the experiences that are the basis for doctrine."[67]

Clausewitz used this same approach in his description of the "critical approach." He wrote that there were three different intellectual courses of action in the critical approach process,

> First, the discovery and interpretation of equivocal facts. This is historical research proper, and has nothing in common with theory.
> Second, the tracing of effects back to their causes. This is critical analysis proper. It is essential for theory; for whatever in theory is to be defined, supported. or simply described by reference to experience can only be dealt with in this manner.
> Third, the investigation and evaluation of means employed. This last is criticism proper, involving praise and censure. Here theory serves history, or rather the lessons to be drawn from history.[68]

Efficacy of the Principles of War

There is no consensus among theorists or practitioners on the proper application of the principles of war or, for that matter, what the correct principles are, how many there are, or how they should be applied. Many writers, such as Jomini, formulated them to be used mainly in tactical situations, often employing the principles as inflexible rules of engagement. (Jomini believed that war was not merely a state of confusion but a phenomenon where leaders such as Napoleon could use principles that had always been applicable in war to bring about success.[69]) Others, such as Clausewitz, took a more liberal position, as can be ascertained from his discussion on theory:

> It [theory] is an analytical investigation leading to a close *acquaintance* with the subject; applied to experience, in our case, to military history, it leads to thorough *familiarity* with it. . . . Theory

---

[67] Ibid., 173-174.
[68] Clausewitz, 156.
[69] Richard A. Preston and Sydney F. Wise, *Men in Arms, A History of Warfare and Its Interrelationships with Western Society*, 4th ed. (New York: Holt, Rinehart and Winston, 1978), 208.

will have fulfilled its main task when it is used to analyze the constituent elements of war, to distinguish precisely what at first sight seems fused, to explain in full the properties of the means employed and to show their probable effects, to define clearly the nature of the ends in view, and to illuminate all phases of warfare in a thorough critical inquiry.... Theory exists so that one need not start afresh each time sorting out the material and plowing through it, but will find it ready to hand and in good order.[70]

But then Clausewitz limits the utility of theory in practice:

If the theorist's studies automatically result in principles and rules, and if truth spontaneously crystallizes into these forms, theory will not resist this natural tendency of the mind. On the contrary, where the arch of truth culminates in such a keystone, this tendency will be underlined. But this is simply in accordance with the scientific law of reason, to indicate the point at which all lines converge, but never to construct an algebraic formula for use on the battlefield. Even these principles and rules are intended to provide a thinking man with a frame of reference for the movements he has been trained to carry out, rather than to serve as a guide which at the moment of action lays down precisely the path he must take.[71]

But even Clausewitz would sanction the use of principles when applied in the proper context, and he did, in fact, develop two principles of his own: "The first principle is: act with the utmost concentration. The second principle is: act with the utmost speed."[72]

Even before Clausewitz penned his theories, however, Niccolò Machiavelli addressed the rational approach to statecraft and the desirability of principles in the use of military force. "We see that Machiaevelli tried to base his science on a number of general postulates."[73]

Napoleon, one of the most successful practitioners of warfare in history, gave much credit to the principles of war. He identified them as those which:

Have regulated the great captains whose deeds have been handed down to us by history: Alexander, Hannibal, Caesar, Gustavus Adolphus, Turenne, Prince Eugene and Frederick the Great. The history of [their] campaigns, carefully written, would be a complete

---

[70] Clausewitz, 141.

[71] Ibid., 141.

[72] Clausewitz, 617.

[73] J. Bronowski and Bruce Mazlish, *The Western Intellectual Tradition* (New York: Harper & Row, Publishers, 1975), 32.

treatise on the art of war; the principles which ought to be followed in offensive and defensive war, would flow from it spontaneously.[74]

Other military theorists are more definitive in their belief that the principles of war have and will play a significant role in the application of strategy. Jomini is probably the most closely associated with the principles of war based on his book, *The Art of War*. In it, his principles are found primarily in his discussion on tactics and maneuver; while his major contribution to strategic thought can be confined to two areas. First, he was in accord with most other writers of his time in the primacy of the state over the military: "The first care of its [the army's] commander should be to agree with the head of the state upon the character of the war."[75] Second, his *fundamental* principle of war employed strategic movement to mass the army "upon the decisive points of a theater of war,"[76] paralleling Clausewitz's concept of the enemy's "center of gravity."

Although identified with employing principles of war as inflexible rules, Jomini qualified his position by recognizing the importance of the commander, thereby tempering his view on the application of unbending maxims. He wrote:

> Of all the theories on the art of war, the only reasonable one is that which, founded upon the study of military history, admits a certain number of regulating principles, but leaves to natural genius the greatest part in the general conduct of a war without trammeling it with exclusive rules.[77]

A.T. Mahan discussed the concept of principles of war in relation to sea power: "While many of the conditions of war vary from age to age with the progress of weapons, there are certain teachings in the school of history which remain constant, and being, therefore, of universal application, can be elevated to the rank of general principles."[78]

---

[74] Napoleon Bonaparte, *Memoirs*, Vol. II (London: H. Colburn and Co., 1823-24), 2; quoted in John M. Collins, *Grand Strategy, Principles and Practices* (Annapolis, MD: Naval Institute Press, 1973), 22.

[75] Jomini, 66.

[76] Ibid., 70.

[77] Antoine Henri Jomini, quoted in J.D. Little, ed., "Jomini and His Summary of the Art of War," in *Military Strategy: Theory and Application*, ed. Colonel Arthur F. Lykke, Jr. (Ret.), (Carlisle Barracks, PA: U.S. Army War College, 1989), 105.

[78] A.T. Mahan, *The Influence of Sea Power Upon History, 1660-1783* 5th ed. (New York: Dover Publications, Inc., 1987), 2.

Liddell Hart also preferred a favorable view toward the principles of war. He listed eight "maxims"—six positive and two negative—that he believed applied to tactics and strategy alike. Those maxims are:

Positive

- Adjust your end to your means.
- Keep your objective always in mind, while adapting your plan to circumstances.
- Choose the line (or course) of least expectation.
- Exploit the line of least resistance—so long as it can lead you to any objective which would contribute to your underlying objective.
- Take a line of operation which offers alternative objectives.
- Ensure that both plan and dispositions are flexible, adaptable to circumstances.

Negative

- Do not throw your weight into a stroke whilst your opponent is on guard.
- Do not renew an attack along the same line (or in the same form) after it has once failed.[79]

Richard Preston and Sydney Wise helped place the study of the principles of war in its rightful place. They maintained that:

> Within the last half century, soldiers and military historians have discovered and enumerated certain basic precepts of condition which operate in warfare and which have regularly affected the decision in past conflicts. These precepts appear to be permanent despite changing conditions. . . . Study of military history in terms of these principles helps to draw attention of the student [and decision maker] to those aspects of a very complicated picture that can help him absorb the many lessons history can teach.[80]

Bernard Brodie took a less positive approach to the principles of war, but did not reject them outright as theory:

> It may well be that the consideration of a catalog of numbered principles (usually fewer than a dozen) with the barest definition of

---

[79] Hart, 335-336.
[80] Preston and Wise, 2.

the meaning of each may be necessary to communicate to second-order minds (or minds too busy with the execution of plans to worry much about the specific validity of the ideas behind them) some conception of what the business is all about. Or it may help the ordinary commander to avoid the most glaring or commonplace errors; just as "the principle of concentration," for example, may help him to refrain from *unjustified* dispersions of his force. However, the commander who is capable of recognizing just as clearly the unique qualities of the situation before him as he does its likeness to somewhat similar situations covered in the textbooks will use such "principles" at most as a reminder of the obvious.[81]

A number of military theorists are critical of using the principles of war as a means for developing strategy or employing military force. One contemporary writer and outspoken critic, Rear Admiral Joseph Wylie, stated:

I think that what the principles really are is an attempt to rationalize and categorize common sense. I am not at all sure whether it is either necessary or possible to teach common sense; but I am very sure indeed that the subject of strategic analysis and understanding is not to be coped with by any such elemental and facile tabulations as these. At the risk of treading on the toes of sincere and able men, I suggest that worship of any such pattern as the "principles of war" is an unaware substitution of slogan for thought, probably brought about by the intellectual formlessness that must inevitably exist when there is no orderly and disciplined pattern of fundamental theory from which one consciously or unconsciously takes departure.[82]

In the last foreign policy address of his administration, former President George Bush voiced his concern of the use of immutable rules when considering the employment of military force for the furtherance of U.S. foreign policy. In a speech before the Corps of Cadets at West Point he warned:

In the complex new world we are entering, there can be no single or simple set of fixed rules for using force. Inevitably, the question of military intervention requires judgment; each and every case is unique. To adopt rigid criteria would guarantee mistakes involving American interests and American lives. And it would give would-be troublemakers a blueprint for determining their own actions; it could

---

[81] Brodie, *War & Politics*, 448.
[82] Wylie, 20.

signal U.S. friends and allies that our support was not to be counted on.[83]

Although he did not take as strong a position as Wylie did against using set principles, he certainly made his beliefs clear on how *not* to conduct strategic policy. So, how should one use the principles of war as a framework in the decision-making process without those same principles becoming "checklist" items that would invariably fail under many given situations?

---

[83] George H. Bush, "U.S. Military Power Must Help Promote Peace," Speech presented at the United States Military Academy at West Point on 5 January 1993.

The Principles of War

Militaries in many nations have principles of war that diverge significantly from those outlined in U.S. military doctrine. (See Table 3.1.)

United States	Great Britain Australia	Soviet Union	People's Republic of China	South Africa [a]
Objective	Selection & Maintenance of AIM		Selection & Maintenance of AIM	Selection & Maintenance of AIM
Offensive	Offensive Action		Offensive Action	Offensive Action
Mass	Concentration of Force	Massing & Correlation of Force	Concentration of Force	Concentration of Force
Economy of Force	Economy of Force	Economy/ Sufficiency of Force		Economy of Force
Maneuver	Flexibility	Initiative	Initiative & Flexibility	Maneuver
Unity of Command	Cooperation		Coordination	Unity of Command
Security	Security			Security
Surprise	Surprise	Surprise	Surprise	Surprise
Simplicity				
	Maintenance of Morale	Mobility & Tempo Simultaneous Attack Preservation of Combat Effectiveness Interworking & Coordination	Morale Mobility Political Mobilization Freedom of Action	Maintenance of Morale Flexibility Cooperation Logistic Support Maintaining of Reserves

Table 3.1 Principles of War

For analytical purposes, however, those basic principles initially identified by J.F.C. Fuller after World War I and later adapted by the United States military will provide the framework.

Of course, volumes have been written by classical and contemporary military theorists alike describing these principles; however, only limited references will be presented to illustrate the essence of each. Following are the nine principles as outlined in *JCS Pub 3-0: Doctrine for Unified and Joint Operations*, briefly described and analyzed.

Objective: Direct every operation toward the achievement of an objective that is clearly defined, attainable, and decisive.

> No one starts a war—or rather, no one in his senses ought to do so—without first being clear in his mind what he intends to achieve by that war and how he intends to conduct it. The former is its political purpose; the latter its operational objective.

Clausewitz placed the principle of the objective in perfect perspective and underscored the crux of the problem in the decision-making seam between the civilian leadership and the military chain of command at the strategic level of war when he wrote:

> The first, the supreme, the most far-reaching act of judgment that the statesman and the commander have to make is to establish . . . the kind of war on which they are embarking; neither mistaking it for, nor trying to turn it into, something that is alien to its nature. This is the first of all strategic questions and the most comprehensive.[84]

The end result is that the political purpose and the military operational objective must complement one another if there is going to be any hope of successfully countering or deterring a crisis situation. Clausewitz stated that "the political object is the goal, war is a means of reaching it, and means can never be considered in isolation from their purpose."[85] He then went on to further define the relationship between the political aim and military objective, "The political object—the original motive for the war—will thus determine both the military objective to be reached and the amount of effort it requires."[86]

Clausewitz focused on the "engagement" in his discussions on strategy, so he carried this notion one step further in his discussion on the objective. He believed that "of all the possible aims in war, the destruction of the enemy's armed forces always appears as the highest."[87]

Needless to say, determining the political aim, the military objective, and formulating a plan for the application of force cannot be performed in isolation—the threat must be considered at the same time. Peter Paret paraphrased Clausewitz's position as follows: "The military objective is dependent on the political purpose, but also on the enemy's political and military policies, and on the conditions and

---

[84] Clausewitz, 86.
[85] Ibid, 87.
[86] Ibid, 81.
[87] Ibid, 99.

resources of the two antagonists, and should be proportionate to these factors."[88]

Although Liddell Hart was often quite critical of Clausewitz, he did support Clausewitz's position that there is a close relationship between the military and political objectives. Hart wrote in *Strategy* that:

> In discussing the subject of "the objective" in war it is essential to be clear about, and to keep clear in our minds, the distinction between the political and the military objective. The two are different but not separate. . . . The military objective is only the means to a political end. Hence the military objective should be governed by the political objective, subject to the basic condition that policy does not demand what is militarily— that, is practically—impossible.
>
> Thus any study of the problem ought to begin and end with the question of policy.[89]

Once the political and military objectives are melded into a synergetic strategy for dealing with a crisis, leadership intent must be communicated to the operational commander(s). World War 11 affords two excellent examples of how strategic objectives were translated from policy to successful results on the battlefield.

Perhaps the best known directive conveying the objective for a military campaign was the one transmitted from the Combined Chiefs of Staff in Washington, D.C. to General Eisenhower as he prepared for the Allied invasion of Europe: "You will enter the continent of Europe and, in conjunction with the other Allied Nations, **undertake operations aimed at the heart of Germany and the destruction of her Armed Forces** [emphasis added]."[90] When commenting on this directive, General Eisenhower professed that "this purpose of destroying enemy forces was always our guiding principle."[91]

The outline campaign plan for "Granite II" provides another good World War II example, this one from the Pacific Ocean Theater. The plan was used to promulgate Admiral C.W. Nimitz's concept of operations to his component commanders for offensives against the

---

[88] Peter Paret, ed., *Makers of Modern Strategy From Machiavelli to the Nuclear Age* (Princeton: Pnnceton University Press, 1986), 207.

[89] Hart, 338.

[90] Dwight D. Eisenhower, *Crusade in Europe* (New York: Da Capo Press, 1977), 225.

[91] Ibid., 225.

Japanese during 1944 and early 1945. In accordance with a strategic concept approved by the Combined Chiefs of Staff, the objective of Granite II was,

> To obtain positions from which the ultimate surrender of JAPAN can be forced by intensive air bombardment, by sea and air blockade, and by invasion if necessary. The ultimate strategic objective is to establish our sea and air power, and if necessary our amphibious forces, in those positions and force the unconditional surrender of JAPAN. [emphasis added][92]

The above discussion and examples appear to be logical and self-evident. They are; yet U.S. decision makers have a tendency at times to lose sight of the true objective. The primary modus operandi used prior to the advent of nuclear weapons was for the political leadership to allow the military to fight the war with little or no supervision once the battle was joined. But in contemporary times this "hands off" approach has changed dramatically.

> In the nuclear era, then, control has become the essence of policy and strategy. For the superpower this has meant the selection of limited political objectives which have a high probability of being accomplished; policies that avoid the direct confrontation of the military forces of the other super power; and strategies that apply limited military means carefully and in a constrained manner.[93]

The most notable example of this constraint on the conduct of war and limited military objectives was the Vietnam War (which has already been addressed) and the change in strategic doctrine leading up to it.

In the late 1950s there was a push to provide options for "a doctrine which left no room for intermediate positions between total peace and total war,"[94] but some believed that the President should be given latitude for options between the two extremes. Unfortunately, what was originally proposed in the early 1960s as a means of achieving objectives short of total war through the use of options addressing limited war became an end in itself. Henry Kissinger, for example, believed that options employing a "graduated response" strategy would

---

[92] Ibid, 228.

[93] Keith A. Dunn, "The Missing Link in Conflict Termination Thought: Strategy," in *Conflict Termination and Military Strategy Coercion, Persuasion, and War*, eds. Stephen J. Cimbala and Keith A. Dunn (Boulder, CO: Westview Press, 1987), 177.

[94] Henry A. Kissinger, *Nuclear Weapons and Foreign Policy* (New York: Harper & Row, 1957), 29.

allow the United States to win limited victories without having to resort to general nuclear war. This strategy was applied in Vietnam but was, regrettably, employed as a military response to escalating North Vietnamese activity rather than being used to establish an allied political objective. In effect, the allied forces were always one step behind the enemy, and the overall effort lost sight of the final political purpose. Stephen Cimbala insightfully described how U.S. intervention in Vietnam "was marked by confusion at the highest policy levels about our objectives, including our objectives for war termination. In the end 'Vietnamization' became a synonym for neither winning nor losing, but merely degrading gracefully."[95]

Landing U.S. Marines in Beirut, Lebanon, in 1982 was another case in point. The Marines were placed there without a definitive objective as to what they were supposed to do, how they were supposed to do it, what was the final objective was, and when they were supposed to leave. In the end, the mission became bogged down in a no-win situation and muddled along until disaster struck.

The Vietnam War was not the only conflict in recent U.S. history that was waged without distinct, identifiable objectives. The Korean War provided another excellent example of where the failure to identify well-defined objectives allowed a conflict to diverge from its original purpose. The FY 1987 Annual Defense Report described the problem as follows:

> Leaving the objective undefined invites an escalation of ambitions in response to battlefield successes. In Korea, we paid the full costs of this lesson—though this lesson still eludes many. There, our original purpose had been to defeat North Korean aggression and restore South Korea's territorial integrity. But as we accomplished that objective, our failure to be entirely clear about what we were fighting for left us vulnerable to the entreaties of those who wanted more. Without adequate assessment of the risks and costs, we crossed the 38th Parallel and pushed forward to the Chinese border, provoking Chinese intervention, multiplying our losses, and eventually leading to a stalemate at the 38th Parallel.[96]

---

[95] Stephen J. Cimbala and Keith A. Dunn, eds., *Conflict Termination and Military Strategy: Coercion, Persuasion, and War* (Boulder, CO: Westview Press, 1987), 3.

[96] Report of the Secretary of Defense Caspar W. Weinberger to the Congress on the FY 1987 Budget, FY 1988 Authorization Request and FY 1987-91 Defense Programs (Washington, DC: U.S. Government Printing Office, 1986), 273, quoted in Cimbala and Dunn, 181.

Offensive: Seize and exploit the initiative to set the terms of the engagement. Military victory requires decisive use of offensive action, often in coordination with defensive action. The aim is to generate an operational momentum to which enemies cannot successfully react, depriving them of freedom of action.

Sun Tzu believed that the offensive is of paramount importance, even if one is put on the defensive. When asked what an army should do when a well ordered army is about to attack, he presented his position in a very concise manner: "Seize something he cherishes and he will conform to your desires."[97]

Clausewitz believed the defense to be stronger than offensive action but felt the use of the defense should be only a temporary measure.

> If defense is the stronger form of war, yet has a negative object, it follows that it should be used only so long as weakness compels, and be abandoned as soon as we are strong enough to pursue a positive object. When one has used defensive measures successfully, a more favorable balance of strength is usually created; thus, the natural course in war is to begin defensively and end by attacking.[98]

Clausewitz referred indirectly to the offensive with respect to coercion in the negotiating process. He stated: "If the enemy is to be coerced you must put him in a situation that is even more unpleasant than the sacrifice you call on him to make. The hardships of that situation must not of course be merely transient, at least not in appearance."[99] In most cases, the offensive should be maintained even when negotiations to end the war are being considered or are currently underway. It is incredible how the United States lost sight of this important principle during the negotiating process to end the Vietnam War.

Jomini saw the importance of the offensive, but with qualifications.

> For a single operation, which we have called taking the initiative, the offensive is almost always advantageous, particularly in strategy. Indeed, if the art of war consists in throwing the masses upon the decisive points, to do this it will be necessary to take the initiative . . . .

---

[97] Sun Tzu, 134.
[98] Clausewitz, 358.
[99] Ibid., 77.

> A defensive war is not without its advantages, when wisely conducted. It may be passive or active, taking the offensive at times.[100]

Jomini's position on the advantages of the offense stemmed from what he considered to be the great principle underlying all operations of war, to include mass, maneuver, and economy of force. He called this the "fundamental principle of war," with this principle embodied in the following maxims:

> 1. To throw by strategic movements the mass of an army, successively, upon the decisive points of a theater of war, and also upon the communications of the enemy as much as possible without compromising one's own.
>
> 2. To maneuver to engage fractions of the hostile army with the bulk of one's forces.
>
> 3. On the battle-field [sic], to throw the mass of the forces upon the decisive point, or upon that portion of the hostile line which it is of the first importance to overthrow.
>
> 4. To so arrange that these masses shall not only be thrown upon the decisive point, but that they shall engage at the proper times and with energy.[101]

While most strategic thinkers generally agree on the primacy of the offense over the defense, Giulio Douhet was unabashedly in favor of the attack. His theory on warfare in the air was essentially the unrestrained offensive.

> Viewed in its true light, aerial warfare admits of no defense, only offense. We must resign ourselves to the offensives the enemy inflicts upon us, while striving to put all our resources to work to inflict even heavier ones upon him. This is the basic principle which must govern the development of aerial warfare.[102]

---

[100] Jomini, 72-73.

[101] Ibid., 70.

[102] Giulio Douhet, *The Command of the Air*, translated by Dino Ferrari (New York: Coward-McCann, Inc., 1942; reprint, Washington, DC: Office of Air Force History, 1983), 55.

Mass: Concentrate sufficient combat power at the correct time and place to achieve decisive results. At the same time, force the enemy to dissipate its strength so that it cannot concentrate.

Clausewitz considered the principle of mass to be important but not a requisite for victory. He wrote, "The best strategy is always to be very strong; first in general, and then at the decisive point . . . . there is no higher and simpler law of strategy than that of keeping one's forces concentrated."[103] But, as with most of the other principles Clausewitz addressed, he must add a qualification to his general statement: "Superiority of numbers in a given engagement is only one of the factors that determines victory. Superior numbers, far from contributing everything, or even a substantial part, to victory, may actually be contributing very little, depending on the circumstances."[104]

General Billy Mitchell favored the use of aircraft in mass, and during World War I his position was proven correct when over 1,000 airplanes were used successfully in an attack on Saint-Mihiel, France, in 1918. In what would be an uncanny foretelling of the future, to include air operations in World War II and Vietnam, Mitchell wrote,

> The air force rises into the air in great masses of airplanes. Future contests will see hundreds of them in one formation . . . . If we attack a city or locality, we send airplanes over it at various altitudes from two or three hundred feet up to thirty thousand all attacking at once so that if any means of defense were devised which could hit airplanes or cause them to be destroyed from the ground, the efforts would be completely nullified, because they [the enemy] could neither see, hear, nor feel all of them.[105]

In 1921 Douhet even went so far as to suggest that massed bombers were capable of winning wars independently of the other services when he asserted, "Thus the Independent Air Force is shown to be the best way to assure victory, regardless of any other circumstances whatever, when it has been organized in a way suitable to winning the struggle for the command of the air and to exploiting the command with adequate forces."[106] His first principle for governing the air force's

---

[103] Clausewitz, 204.
[104] Ibid., 204.
[105] William Mitchell, *Winged Defense* (New York and London: G.P. Putman's Sons, 1925), 164; quoted in Barry D. Watts, *The Foundations of US Air Doctrine. The Problem of Friction in War* (Maxwell Air Force Base, AL: Air University Press, December 1984), 10.
[106] Douhet, 98.

operations was "An Independent Air Force should always operate in mass."[107] His follow-on principle better defined the first with respect to mass by stating, "Inflict the greatest damage in the shortest possible time."[108]

Hart viewed mass in terms of space and time:

> An army should always be so distributed that its parts can aid each other and combine to produce the maximum *possible* concentration of force at one place, while the minimum force *necessary is* used elsewhere to prepare the success of the concentration....
> Superior weight at the intended decisive point does not suffice unless that point cannot be reinforced *in time* by the opponent.[109]

Economy of Force: Allocate minimum essential combat power to secondary efforts in order to dissipate enemy strength and to achieve superiority in the area where decision is sought.

Clausewitz viewed economy of force from two perspectives—the actual employment of armed forces and the selection of an appropriate military objective. He first of all believed that the commander should ensure that his forces are always involved; that they never be allowed to remain idle while the fight is being waged.

> If a segment of one's force is located where it is not sufficiently busy with the enemy, or if troops are on the march—that is, idle while the enemy is fighting, then these forces are being managed uneconomically . . . Even the least appropriate task will occupy some of the enemy's forces and reduce his overall strength, while completely inactive troops are neutralized for the time being.[110]

Second, the object or goal of using military force economically should be directed at the enemy's center of gravity, "the hub of all power and movement, on which everything depends. That is the point against which all our energies should be directed."[111]

---

[107] Ibid., 49.
[108] Ibid., 51.
[109] Hart, 355.
[110] Clausewitz, 213.
[111] Ibid., 595-596.

Victory in the true sense implies that the state of peace, and of one's people, is better after the war than before. Victory in this sense is only possible if a quick result can be gained or if a long effort can be economically proportioned to the national resources....

The experience of history brings ample evidence that the downfall of civilized states tends to come not from the direct assaults of foes but from internal decay, combined with the consequences of exhaustion in war.[112]

Maneuver: Place the enemy in a position of disadvantage through the flexible use of combat power.

Maneuver is action. It involves positioning military forces and logistics in the field and then employing that combat force in a dynamic means against the enemy. John Collins described it as giving "form and shape to concentration. It is the antithesis of mental stagnation or static physical positions."[113]

Sun Tzu appreciated maneuver's importance to strategy when he wrote, "He who knows the art of the direct and the indirect approach will be victorious. Such is the art of maneuvering."[114] But he also recognized that: "Nothing is more difficult than the art of maneuver. What is difficult . . . is to make the devious route the most direct and to turn misfortune to advantage."[115]

Clausewitz correctly maintained that maneuver was one of those applications of strategic theory where no rules applied in the determination of victory. He wrote:

No rules of any kind exist for maneuver, and no method or general principle can determine the value of the action; rather, superior application, precision, order, discipline, and fear will find the means to achieve palpable advantage in the most singular and minute circumstances. It is on these qualities that victory in this type of contest largely depends.[116]

He also pointed out that "the end for which a soldier is recruited, clothed, armed, and trained, the whole object of his sleeping, eating,

---

[112] Hart, 357-359.

[113] John M. Collins, *Grand Strategy, Principles and Practices* (Annapolis, MD: Naval Institute Press, 1973), 26.

[114] Sun Tzu, 106.

[115] Ibid., 102.

[116] Clausewitz, 542.

drinking, and marching *is simply that he should fight at the right place at the right time.*"[117]

Jomini looked at maneuver as a potential force multiplier:

> The system of rapid and continuous marches multiplies the effect of an army, and at the same time neutralizes a great part of that of the enemy's, and is often sufficient to insure success; but its effect will be quintupled if the marches be skillfully directed upon the decisive strategic points of the zone of operations, where the severest blows to the enemy can be given.[118]

As so often happens in the prosecution of war, basic theories and/or principles tend to be forgotten or ignored in the heat of battle. This was true of the principle of maneuver during the First World War as static lines and trench warfare became the norm. Two strategic writers saw the folly of this stratagem, and after the war attempted to rectify this devastating lapse in military theory. One, J.F.C. Fuller, reintroduced the principle of maneuver by advocating mechanized warfare, with the use of tanks and mobility to break through enemy lines.[119] Liddell Hart took the principle one step further by introducing the concept of the "indirect approach."[120]

An Army officer writing in *Military Review* chose a more general perspective in describing the principle of maneuver. "Maneuver warfare," he wrote, "is a thought process, not a particular set of tactics or techniques. It is based on a firm preference to trap the enemy rather than merely push him away. This thought process influences the prosecution of combat along the entire structure of warfare."[121]

Unity of Command: For every objective, ensure unity of effort under one responsible commander. Other components of unity of effort are common objectives, coordinated planning, and trust.

---

[117] Ibid., 95.

[118] Jomini, 176.

[119] Brian Bond and Martin Alexander, "Liddell Hart and De Gaulle: The Doctrines of Liability and Mobile Defense," in *Makers of Modern Strategy From Machiavelli to the Nuclear Age*, ed. Peter Paret (Princeton: Princeton University Press, 1986), 601.

[120] John Shy, "Jomini," in *Makers of Modern Strategy From Machiavelli to the Nuclear Age*, ed. Peter Paret (Princeton: Princeton University Press, 1986), 181.

[121] John F. Antal, Major, US Army, "Maneuver Versus Attrition, A Historical Perspective," *Military Review* 10 (October 1992): 22.

One of the first military theorists to consider the relationship between the head of state and the military commander, Machiavelli believed that command must be placed in the hands of one man. If the state is a monarchy, the ruler himself should be the commanding general, he contended, but if the state is a republic, the army in wartime should be entrusted to one commander who should have unlimited authority.[122]

In his description of the principle of "unity," Collins discussed the need to properly focus power. Unity, Collins wrote, "embraces solidarity of purpose, effort and command. It directs all energies, effort, and activities, physical and mental, toward desired ends."[123] He believed a limited amount of coordination and cooperation would help enhance unity of effort when a single chain of command was not possible, but individualism and opportunism dictated singular command if at all possible.

Operations during World War II underlined the need for unity of command as the United States joined forces with other countries to contest the war. This experience resulted in the enactment of the National Security Act of 1947. This act provided legislation for the creation of unified commands to integrate U.S. military land, naval and air forces into a single command structure when called upon to defend the country. History has shown that when applied properly the concept will work (i.e., Persian Gulf War), whereas when the concept is improperly designed and/or employed it will fail (i.e., Vietnam, Desert One, Beirut).

Security: Protect friendly forces and their operations from enemy actions that could provide the enemy with unexpected advantage.

The objective of security is to prevent the enemy from achieving an *unexpected* advantage. To accomplish this task, intelligence operations are employed to take continuous, positive action to prevent surprise, to retain flexibility, and to preserve freedom of action. Technology and the rapid pace of combat today make good intelligence an imperative, but it has not always been recognized in the writings of military theorists.

---

[122] Felix Gilbert, "Machiavelli: The Renaissance of the Art of War," in *Makers of Modern Strategy from Machiavelli to the Nuclear Age*, ed. Peter Paret (Princeton: Princeton University Press, 1986), 25.

[123] Collins, 28.

Clausewitz, for one, often appeared skeptical of the value of intelligence, yet at other times seemed to accept its virtue in a roundabout way. For instance, he noted the importance of surprising the enemy, which, in effect, amounts to a failure of the enemy's intelligence. One could also argue that Clausewitz would postulate that poor intelligence leads to uncertainty, thereby compounding the consequences of friction in war.

Supposition aside, however, Clausewitz wrote pointedly of his disdain for intelligence. He stated in no uncertain terms that "Many intelligence reports in war are contradictory; even more are false, and most are uncertain. . . . In short, most intelligence is false, and the effect of fear is to multiply lies and inaccuracies."[124] He therefore believed that in the conduct of war additional information is of little value.

> During an operation decisions have usually to be made at once: there may be no time to review the situation or even to think it through. Usually, of course, new information and reevaluation are enough to make us give up our intentions: they only call them into question. We know more, but this makes us more, not less uncertain.[125]

At the strategic level he subscribed to the fact that intelligence formed the basis of war plans and operations, but then went on to state that "If we consider the actual basis of this information, how unreliable and transient it is, we soon realize that war is a flimsy structure that can easily collapse and bury us in its ruins."[126]

Sun Tzu, on the other hand, believed that intelligence was critical to the conduct of war: "Know the enemy and know yourself; in a hundred battles you will never be in peril. When you are ignorant of the enemy but know yourself, your chances of winning or losing are equal."[127] He contended that not only must we know who we are fighting and have knowledge of the enemy's plan, but we must also understand the conditions under which we must fight:

> To estimate the enemy situation and to calculate distances and the degree of difficulty of the terrain so as to control victory are virtues of

---

[124] Clausewitz, 117.
[125] Ibid., 102.
[126] Ibid., 117.
[127] Sun Tzu, 84.

the superior general. He who fights with full knowledge of these factors is certain to win; he who does not will surely be defeated.[128]

Surprise: Take action against enemies at times, places, and in manners for which they neither are prepared nor expect.

"All warfare is based on deception,"[129] emphasized Sun Tzu in *The Art of War*. "When capable, feign incapacity; when active, inactivity. When near, make it appear that you are far away; when far away, that you are near."[130]

Clausewitz believed that surprise was basic to all operations because "without it superiority at the decisive point is hardly conceivable."[131] He did caution, however, that surprise had its limits to achieving a successful outcome in battle:

> But while the wish to achieve surprise is common and, indeed, indispensable, and while it is true that it will never be completely ineffective, it is equally true that by its very nature surprise can rarely be *outstandingly* successful. It would be a mistake, therefore, to regard surprise as a key element of success in war. The principle is highly attractive in theory, but in practice it is often held up by the friction of the whole machine.[132]

Clausewitz went on to declare that surprise is more applicable at the tactical than at the operational and strategic levels of war because of time and distance.

> Basically surprise is a tactical device, simply because in tactics time and space are limited in scale. Therefore in strategy surprise becomes more feasible the closer it occurs to the tactical realm, and more difficult, the more it approaches the higher levels of policy.[133]

He did not consider strategic surprise to be a viable option because of the length of time necessary to prepare an army for war. In fact, he deemed it very rare "that one state surprises another, either by an attack or by preparations for war."[134] Of course, he did not have the benefit of

---

[128] Ibid., 128.
[129] Ibid., 66.
[130] Ibid., 66.
[131] Clausewitz, 198.
[132] Ibid., 198.
[133] Ibid., 198.
[134] Ibid., 199.

observing the German blitzkrieg in 1939, Pearl Harbor, Korea, or even Iraq's invasion of Kuwait.

Simplicity: Issue clear, concise, uncomplicated plans, orders, and/or guidance.

Clausewitz did not address simplicity per se, but he did recognize the need for it. He cautioned,

> Everything in war is very simple, but the simplest thing is difficult. The difficulties accumulate and end by producing a kind of friction that is inconceivable unless one has experienced war. . . . Countless minor incidents, the kind you can never really foresee, combine to lower the general level of performance, so that one always falls short of the intended goal.[135]

He placed the onus of overcoming this friction squarely on the shoulders of the general in command of the army, declaring:

> An understanding of friction is a large part of that much-admired sense of warfare which a good general is supposed to possess . . . . The good general must know friction in order to overcome it whenever possible, and in order not to expect a standard of achievement in his operations which this very friction makes impossible.[136]

Clausewitz then emphasized the need for practice and experience to overcome the detrimental effects of this friction and to permit the officer to make the right decisions.

Levels of War

As the study of strategy and doctrine came back into vogue in the early 1980s, one of many major shortcomings identified was the disconnection between the development of strategy at the national level and the employment of tactical operations on the battlefield. Lessons learned from the Vietnam War graphically illustrated this problem. In Vietnam, there was no procedure in place for translating the strategic political aims of the policy makers into strategic military objectives in the operational theater and then developing follow-on plans for the tactical employment of forces. It was not a matter of having never done

---

[135] Ibid., 119.
[136] Ibid., 120.

this type of planning before—the United States did it very well during World War II in both the European and Pacific theaters. It was a case of not codifying a *process* for future reference, a situation that contributed to flawed procedures in Vietnam.

The concept of three "levels of war" was introduced to the U.S. military in Army Field Manual (FM) 100-5, *Operations* in 1982. Although the idea was not new (its roots can be traced back to the Napoleonic Wars and the American Civil War[137]), reference to the practical application of the premise in recent U.S. military literature had been virtually nonexistent.

The 1986 version of FM 100-5 described the levels of war as follows:

> War is a national undertaking which must be coordinated from the highest levels of policy-making to the basic levels of execution. Military strategy, operational art, and tactics are the broad divisions of activity in preparing for and conducting war. Successful strategy achieves national alliance and political aims at the lowest possible cost in lives and treasure. Operational art translates those aims into effective military operations and campaigns. Sound tactics win the battles and engagements which produce successful campaigns and operations.[138]

Figure 3.1 illustrates the fluid relationship between the three levels of war: strategic, operational, and tactical. The overlap in the circles indicates that there are no definitive boundaries between each level where all actions in the conduct of war can be precisely categorized by objective or purpose. The figure also illustrates a correlation between the levels of war and the strategy process.

---

[137] Department of Defense, U.S. Air Force, *Air Force Manual 1-1, Basic Aerospace Doctrine of the United States Air Force*, vol. II (Washington, DC: U.S. Government Printing Office, March 1992), 43.

[138] Department of Defense, Army, *FM 100-5, Operations* (Washington, DC: U.S. Government Printing Office, May 1986), 9.

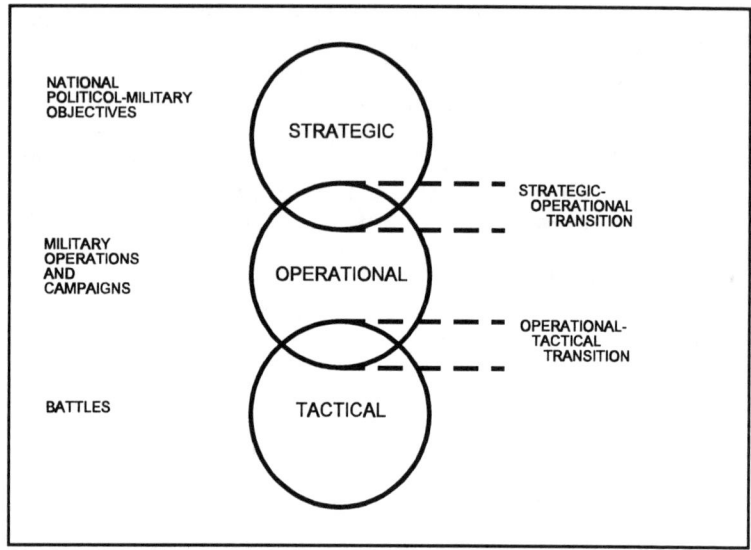

Figure 3.1. Levels of War

Strategic Level

Decision makers develop national security policy, establish objectives, and formulate military strategy in support of that policy at the strategic level. In war, ideally the political goals are defined by the President and the National Command Authority (NCA), and then forwarded to the theater commander, usually through the Chairman of the Joint Chiefs of Staff (CJCS). Although detailed plans for fighting the war are not developed at this level, the significance of the decisions made here cannot be overemphasized. In a study on the conduct of warfare between 1914 and 1945, Allen Millett and Williamson Murray concluded that politico-strategic decisions at the strategic level were generally the key to whether wars were won or lost rather than military operations at the operational or tactical levels.[139]

Operational Level

The operational level of war is concerned with the employment of military force in a theater of operations in order to achieve national objectives. It is at this level that national objectives are coordinated with military objectives on the battlefield through the establishment of

---

[139] Allen R. Millett and Williamson Murray, "Lessons of War," *The National Interest* (Winter 1988-1989: 83-95); quoted in *Basic Aerospace Doctrine*, 44.

strategic military objectives. The operational level is "concerned with employing military forces in a theater of war or theater of operations to obtain an advantage over the enemy and thereby attain strategic military goals through the design, organization, and conduct of campaigns and major operations."[140] FM 100-5 addresses the operational aspects at this level as "operational art," which the manual describes as "involving fundamental decisions about when and where to fight and whether to accept or decline battle. Its essence is the identification of the enemy's operational center-of-gravity, and the concentration of superior combat power against that point to achieve a decisive success."[141]

The plan to deploy troops and equipment to a theater of operations is developed by the theater commander in chief (CINC) in concert with the JCS, who embodies the overlap between the strategic and the operational levels. The employment of military force to and within the theater is the focus of the theater commander and his subordinates at the operational level with the approval of the Chairman, JCS. Directives and movements are orchestrated through the use of a campaign plan developed by the theater commander and his staff.

> Campaigns represent the art of linking battles and engagements in an operational design to accomplish strategic objectives . . . they serve as the *unifying focus* for our conduct of warfare . . . The joint campaign is oriented on the enemy's strategic and operational centers of gravity.[142]

As alluded to earlier, the concept of an operational level of war is not new. Clausewitz understood the significance of the operational level in his discussion on strategy. "A prince or general can best demonstrate his genius by managing a campaign exactly to suit his objectives and his resources, doing neither too much nor too little."[143]

Tactical Level

The tactical level is where maneuver units actually engage the enemy. It "translates potential combat power into success in battles and engagements through decisions and actions that create advantages when in contact with or in proximity to the enemy."[144] The tactical

---

[140] *Basic Aerospace Doctrine*, 46.
[141] *Operations*, 10.
[142] *Joint Warfare*, 45-46.
[143] Clausewitz, 177.
[144] *Basic Aerospace Doctrine*, 47.

level of warfare is not the focus of this book, but tactics do impact the strategic process. In their book, Drew and Snow provided an excellent synopsis of the relationship between tactics and strategy at the higher levels of war:

> *Battlefield strategy* [tactics] is the art and science of employing forces on the battlefield to achieve national security objectives. The classic differentiation between tactics and higher levels of strategy remains relevant in the sense that tactics govern the use of forces on the battlefield while grand strategy, military strategy, and operational strategy bring forces to the battlefield.[145]

War Termination

A government considering a policy of war, or forced with the likelihood of military action, must weigh the prospects, methods, and time associated with ending the crisis in relation to political aim, national will, and available resources. In other words, wars do not end by themselves; there must be a strategy for making them end. The often misunderstood concept of war termination is inherently a political responsibility; the military is only a party to the results.

ICAF identifies two traditional methods for obtaining the objectives of conflict.[146] The first endeavors to dominate or overthrow the opponent's military capability and political policy to reach an imposed settlement (i.e. capitulation). The second seeks concessions, either political, geographic, economic, or military, with the objective being a negotiated settlement rather than an imposed ending.[147]

Capitulation is characterized by one nation attempting to destroy the opponent's ability to resist through the destruction of his forces or disruption of his ability to operate. The objective of the prevailing nation is to impose a settlement of its own choosing. This is the classic approach to war termination sought throughout history. The most celebrated example involving the United States was the unconditional surrender of Germany and Japan that brought about the end of World War II. In Japan's case, the nation was defeated by a combination of the above methods—destruction of her expeditionary

---

[145] Drew and Snow, 20-21.

[146] Industrial College of the Armed Forces, *Military Power and Strategy: Force Determination*, Faculty Guide, Academic Year 1989-1990, 11-2 to 11-5.

[147] This section of war termination theory is based on information contained in James E. Toth, "Winning War and Peace," in *The Soviet Challenge in the 1990s*, ed. Stephen J. Cimbala (New York: Praeger, 1989), 167-183.

forces and naval power in concert with the strangulation of her national economic base that supported the forces. Ending a war through capitulation is more difficult today, however, as the scope of warfare has been magnified through the employment of larger armies, increased destructive capability of weaponry, and expanding geographic range. History bears this out. With the possible exception of the Gulf War, the United States has not encountered a need for war termination through capitulation since the end of World War II.

The objective of a **negotiated** end to war is to gain concessions from the opposing nation. Waging war under these terms can be either offensive- or defensive-oriented, depending upon the political aim. Under both circumstances, however, military and diplomatic efforts must function in consonance, although emphasis may be placed on one or the other at any given time.

The goal of a war waged for concessions is the negotiation of a settlement under advantageous conditions, not domination or dissolution of the opposition. In fact, the continued existence of an economically and politically viable state may be the preferred solution in order to maintain regional peace and stability. The principal objective in this case is to create a peace that achieves the limited war objectives of the prevailing nation while eliminating conditions that could provoke a resumption of hostilities at a later date.

A negotiated end to war can take many forms, the most significant one being the armistice or truce. This differs from a negotiated settlement in that the parties to a truce recognize that it is merely a temporary measure rather than a final solution, a respite to war rather than a return to peace. The theory behind a truce is that it provides the time to negotiate a permanent settlement. The Korean War is an excellent example of hostilities ending in a truce—combat ceased in 1953, and there still has not been a negotiated settlement.

Many factors play a role in war termination irrespective of what method is used to bring about a conclusion. Perhaps the most significant consideration is the outcome of the war relative to the means employed in its conduct and the anticipated gains at the conclusion of hostilities. In World War II the United States was fighting for national survival, which justified economic and personal sacrifice from the populace. During the Vietnam War, on the other hand, no legitimate political goal was articulated by the government with respect to the personal sacrifice being made to sustain the war effort. As a result, public support was lost and the war was doomed to fail.

Classical strategic theorists recognized the significance of war termination and embraced the concept as a prominent part of their writings. Sun Tzu was perhaps the first to address the subject, and many of his theories are still relevant today. Of most importance then and now, he believed in victory as "the main object of war."[148] He later qualified this concept, driving home a sore point with regard to Vietnam: "[W]hat is essential in war is victory, not prolonged operations."[149]

Clausewitz was one of the first classical strategic thinkers to write about the correlation between war termination and political policy. In describing this relationship, he stated:

> Since war is not an act of senseless passion but is controlled by its political object, the value of this object must determine the sacrifices to be made for it in *magnitude* and also in *duration*. Once the expenditure of effort exceeds the value of the political object, the object must be renounced and peace must follow.[150]

It was this close relationship between policy and war that caused Clausewitz to recognize the lack of finality in war termination: "the ultimate outcome of war is not always to be regarded as final. The defeated state often considers the outcome merely as a transitory evil, for which a remedy may still be found in political conditions at some later date."[151]

Clausewitz's influence on Liddell Hart can again be found in Hart's discussion on war termination:

> The object of war is a better state of peace.... Hence it is essential to conduct war with constant regard to the peace you desire....
>
> History shows that gaining military victory is not in itself equivalent to gaining the object of policy.... In consequence, whenever war has broken out, policy has too often been governed by the military aim—and this has been regarded as an end in itself, instead of as merely a means to the end.[152]

Paul Seabury and Angelo Codevilla emphasized one further facet of war termination: the significance of the role governmental policy plays at the conclusion of hostilities, particularly when one side is forced to capitulate. "The degree of severity following a surrender depends on the character of the winning troops, the level of their discipline, the

---

[148] Sun Tzu, 73.
[149] Ibid., 76.
[150] Clausewitz, 92.
[151] Ibid., 80.
[152] Hart, 338.

amount of hatred that has been built up during the war, and, above all the policy pursued by the winner."[153]

---

[153] Paul Seabury and Angelo Codevilla, *War; Ends & Means* (New York: Basic Books, Inc., 1989), 262.

Chapter 4

National Security Policy Process

The national security policy process in the United States is a complex procedure, yet in theory can be described in straightforward, rational terms. The complexity comes in having to coalesce multifarious disciplines integral to the successful implementation of the design.

Understanding the whole national security policy process is integral to comprehending how war becomes an instrument of policy and appreciating the complexity inherent in the decision-making seam between civilian and military leadership (which falls within the development and employment of NATIONAL SECURITY STRATEGY).

One can address the theoretical foundations of national security policy in many ways. One avenue is to use a derivation of the model created by the faculty of the National Defense University to illustrate the framework for developing grand strategy, as portrayed in Figure 4.1.

As presented here, the Framework for Grand Strategy can be seen as merely a "peacetime" model for developing strategy rather than a concept for decision making in a crisis situation. This is certainly appropriate. Grand strategy must, of necessity, be developed in peacetime to guide decision makers in developing national policy and conducting responses to crisis situations. The correct strategy developed during peacetime provides a solid basis to confront aggression. An ill-conceived strategy, or a strategy impracticable due to a lack of resources, places the decision maker at a distinct disadvantage by limiting his options at the very beginning of a crisis situation. An unenviable position to be in, it is one of the primary tangible factors that influence decision making at the seam between civilian and military leadership.

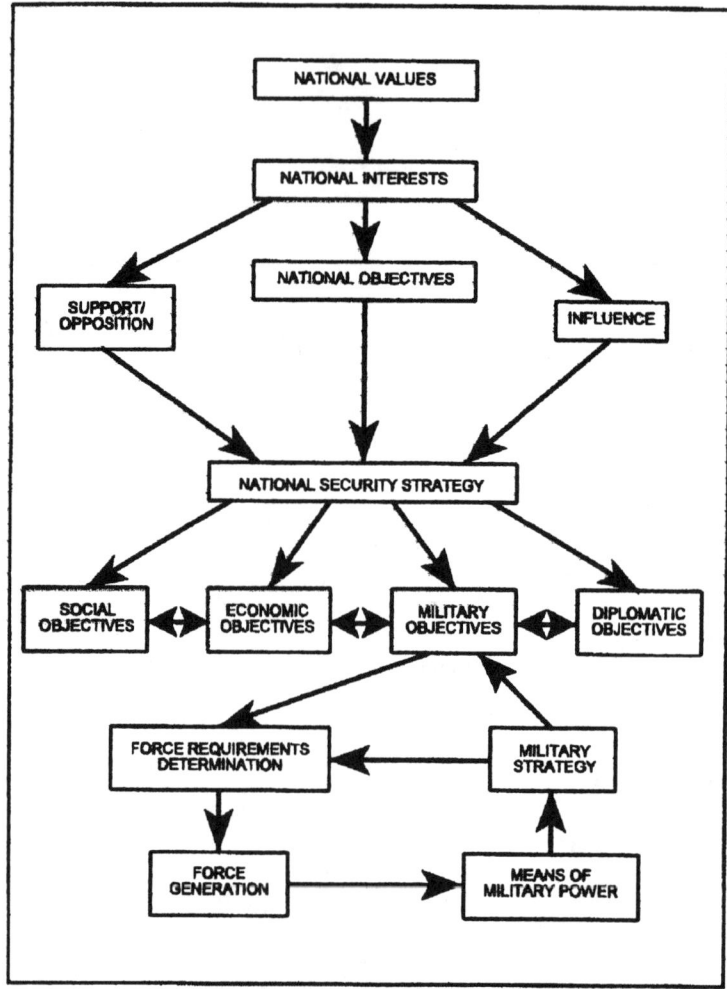

Figure 4.1 Framework for Grand Strategy

Rear Admiral Henry Eccles summarized the relationship as follows: "Modern strategy requires an intuitive synthesis of policy, political purpose, values, military power and force, military readiness and effectiveness, economics, logistics and the process of negotiation."[154] Yet these same concepts also apply when the leadership must deal with

---

[154] Eccles, 41.

National Security Policy Process

a critical decision supporting a crisis situation. Simply put, theoretically and in practice, the national security process reflects four key influences: values, national interests, national security strategy, and military strategy.

- VALUES: Derived from the Declaration of Independence, the Constitution, and heritage, the basic values of life, liberty, and the pursuit of happiness have endured since the birth of this nation. They represent the broad goals that have guided U.S. foreign policy throughout history by forming the basis for the country's national interests. The essence of U.S. military strategy stems directly from the preamble to the Constitution, where its conduct relates directly to American values:

> *We the People* of the United States, in Order to form a more perfect Union, establish Justice, insure domestic Tranquility, provide for the common defence, promote the general Welfare, and secure the Blessings of Liberty to ourselves and our Posterity, do ordain and establish this Constitution for the United States of America.[155]

- NATIONAL INTERESTS: In the March 1990 edition of the *National Security Strategy of the United States*, President Bush outlined America's national interests. Like the values of the American people, national interests have remained fairly constant over time and can be summarized as follows:

  - The survival of the United States as a free and independent nation, with its fundamental values intact and its institutions and people secure.

  - A healthy and growing U.S. economy to ensure opportunity for individual prosperity and a resource base for national endeavors at home and abroad.

  - A stable and secure world, fostering political freedom, human rights, and democratic institutions.

  - Healthy, cooperative and politically vigorous relations with allies and friendly nations.[156]

---

[155] *Constitution*, preamble.
[156] The White House, *National Security Strategy of the United States* (Washington, DC: The White House, March 1990), 2-3.

It is important to look at a subset of national interests, "vital interests," when considering the use of military force in response to a crisis situation. According to Bernard Brodie, our vital interests "are those interests against the infringement of which we are prepared to take some kind of serious military action. [They] exclude many issues of genuine and even important national interest . . . ."[157] But who decides what are vital interests and how they are defined? Does the President have this responsibility? What about Congress? Decisions on whether or not to use military force and the appropriate use of it in the future depend on the answer.

- NATIONAL SECURITY STRATEGY: Derived from the national interests, national security strategy can be divided into four component objectives: social (information), economic, military, and political. Each aim has its own strategy, and in each case the challenge comes in relating means to ends during peacetime.

Overall priorities for the military component of national security strategy, based on the objective of preserving the United States as a free and independent nation, are as follows:

- Safeguarding the United States and its allies and interests by deterring aggression and coercion across the conflict spectrum, and should deterrence fail, by defeating armed aggression and ending hostilities on terms favorable to the United States and its allies.

- Encouraging and assisting our allies and friends in defending themselves against aggression, coercion, subversion, insurgency, terrorism, and drug trafficking.

- Ensuring access to critical resources, markets, the oceans, and space for the United States, its allies, and friends.

- Where possible, neutralizing Soviet military presence and influence throughout the world, increasing the disincentives for Soviet use of subversive force, and encouraging independent policies by Soviet client states.

- Preventing the transfer of military critical technology and resources to the Soviet bloc and hostile countries or groups.

---

[157] Bernard Brodie, *Strategy and National Interests, Reflections for the Future* (New York: National Strategy Information Center, Inc., 1971), 12.

- Preventing the spread of nuclear, chemical, and biological weapons.

- Reducing reliance on nuclear weapons by strengthening conventional and chemical deterrents; pursuing equitable and verifiable arms reduction agreements and insisting on compliance with such agreements; and pursuing technologies for strategic defense.

- Addressing the root causes—military, political, economic and social—of regional instabilities, and maintaining stable regional military balances.[158]

- MILITARY STRATEGY: Until the breakup of the Soviet Union, America's grand strategy had been relatively consistent for 45 years, and could be correlated with one major objective, the *containment* of Soviet expansion. To support this strategy, the military component was required to adjust on occasion in response to a changing threat. Although the U.S. military is currently going through another significant change, this time in response to a lessening threat, U.S. defense policy *still* hinges on the following elements to support the nation's grand strategy:

  - Deterrence: Throughout the postwar period we have deterred aggression and coercion against the United States and its allies by persuading potential adversaries that the costs of aggression, either nuclear or conventional, would exceed any possible gain. "Flexible response" demands that we preserve options for direct defense, the threat of escalation, and the threat of retaliation.

  - Strong Alliances: Shared values and common security interests form the basis of our system of collective security. Collective defense arrangements allow us to combine our economic and military strength, thus lessening the burden on any one country.

  - Forward Defense: In the postwar era, the defense of these shared values and common interests has required the forward presence of significant American military forces in Europe, Asia, and the Pacific, and at sea. These forces provide the capability, with our

---

[158] Department of Defense, *Report of the Secretary of Defense Frank C. Carlucci to the Congress on the FY 1990/FY 1991 Biennial Budget and FY 1990-94 Defense Programs*, by Frank C. Carlucci (Washington, DC: Department of Defense, January 17, 1989), 34.

allies, for early, direct defense against aggression and serve as a visible reminder of our commitment to the common effort.

- Force Projection: Because we have global security interests, we have maintained ready forces in the United States and the means to move them to reinforce our units forward deployed or to project power into areas where we have no permanent presence. For the threat of protracted conflict we have relied on the potential to mobilize the manpower and industrial resources of the country.[159]

These four elements make up the strategic blueprint of what civilian and military leaders have available to them to support decisions in a given crisis. It was this blueprint that formed the foundation to support President Bush's response to the Iraqi invasion of Kuwait.

In addition to the above, during peacetime, FORCE REQUIREMENTS are determined based on military objectives and strategy developed in support of national security strategy. They are agreed on during the annual budget cycle. These requirements, however, have historically fallen short of providing the resources necessary to fully support the various strategies committed to since World War II.[160]

Similarly, FORCE GENERATION capability, which evolves from force requirements determination and involves the development and acquisition of resources, provides the foundation for MILITARY POWER, the definitive force underpinning military strategy.

The United States' National Security Policy Process

When attempting to research the principles of war and integrate them into the decision-making process, one must have a comprehensive understanding of the U.S. military organization and strategic doctrine, including:

- Department of Defense organization
- Military chain of command
- Military strategic doctrine

---

[159] White House, *National Security Strategy of the United States*, 23.

[160] Jeffery Record, *Revising U.S. Military Strategy, Tailoring Means to Ends* (Washington: Pergamon-Brassey's International Defense Publishers, 1984), 11.

Department of Defense Organization

The end of World War II brought with it new global responsibilities for the United States and a new organization for national security. The historic concept of having separate military organizations, a War Department and a Navy Department, was no longer adequate to support increased U.S. political and military roles in the world. The National Defense Act of 1947 was therefore enacted by Congress to create a Department of Defense headed by a civilian secretary. The act also established the Air Force as a separate service, and set up the Joint Chiefs of Staff as a formal, corporate body composed of the three service chiefs.

Although the 1947 act significantly altered the military organization and the way the Services conducted business, amendments were required to help delineate lines of command, authority relationships, and responsibilities for roles, missions, budgets, and resource acquisition. Acts passed by Congress in 1949, 1953, and 1958 improved the national security process by creating the position of Chairman, Joint Chiefs of Staff, giving the Secretary of Defense more authority, and clarifying the chain of command, particularly with respect to unified and specified combatant commands.

Even with these changes, however, problems remained. The structure of the JCS had evolved into an unworkable organization where military advice to the civilian decision makers was often equated with pablum. The Chairman of the JCS was little more than a figurehead for the organization. Commanders of the war-fighting commands were frequently left out of the decision-making process, from both an operational and a resource allocation point of view. The military services not only controlled future weapons procurement but also dominated roles and missions development, a principal source of budgetary authority for each Service. The Secretary of Defense retained only varying influence over the military establishment, depending on the historical time frame and the personality of the Secretary at the time.

Contemporary events graphically illustrated these problems, including: Vietnam, Mayaguez, Desert One (Iran rescue mission), Beirut (1983), Grenada, and the Vincennes affair in the Persian Gulf. Military advice to the President and excessive procurement costs also highlighted problems in the chain of command, military advice to the national command authorities, and the JCS organizational structure itself. Most studies on the need for reorganization identified the strength of the separate Services in pursuing their own basic interests as the main problem with the system.[161]

To address these deficiencies, Congress enacted the Goldwater-Nichols Act of 1986 (also known as the Defense Reorganization Act). Major provisions of the bill relevant to this discussion are as follows. The bill:

- Designated the Chairman of the Joint Chiefs of Staff as the principal military advisor to the President.

- Transferred to the JCS Chairman the principal duties previously performed by the corporate JCS and updated and expanded those duties.

- Specified that the JCS Chairman manages the Joint Staff and prescribes its duties and staffing procedures.

- Required the JCS Chairman to submit a report every three years to the Secretary of Defense on the appropriateness of the roles and missions of the four Services.

- Required the JCS Chairman to prepare fiscally constrained strategic plans.

- Required the JCS Chairman to advise the Secretary of Defense on the extent to which the program recommendations and budget proposals of the Military Departments conform with the priorities established in strategic plans and with the operational requirements of the unified and specified combatant commands.

- Specified that the operational chain of command, unless otherwise directed by the President, runs from the President to the Secretary of Defense to the unified and specified combatant commanders.

- Authorized the President or Secretary of Defense to place the JCS Chairman in the channel of command communications between the Secretary of Defense and the combatant commanders.

---

[161] Vincent Davis, "The Evolution of Central U.S. Defense Management," in *Reorganizing America's Defense, Leadership in War and Peace*, eds. Robert J. Art, Vincent Davis, and Samuel P. Huntington (Washington, DC: Pergamon-Brassey's International Defense Publishers, 1985), 162.

- Authorized the combatant commanders to specify the chains of command and organizational relationships within their commands.

- Strengthened and expanded the "full operational command" authority of combatant commanders.

- Strengthened the authority of the combatant commanders over the selection, retention, and evaluation of their staff members and their subordinate commanders.

- Specified the responsibilities of the Secretaries of the Military Departments to the Secretary of Defense.

- Specified that the functions of the Military Departments (to recruit, organize, supply, equip, train, etc.) are undertaken to meet the operational requirements of the combatant commands.[162]

The Defense Reorganization Act forced profound changes in the way the Department of Defense did (and does) business based primarily on the revised command relationships among the participants in the organization. For example, the authority of the Chairman was strengthened significantly. He not only became the primary advisor to the President and head of the JCS, but also formally became a conduit (at the discretion of the President) for the communications flow to/from the combatant commanders, a spokesman for the combatant commanders in the budget process, and the approval authority for the Services' roles and missions. The combatant commanders' positions as CINCs were also strengthened as they became commanders in fact as well as in name. Additionally, the Services were obliged to address CINC requirements and priorities during the annual budget process. The aforementioned changes were made at the expense of Service autonomy as the Services were placed in more of a support role to the Chairman and the CINCs.

Chain of Command

The JCS identifies the "mission to be accomplished" and the "objective to be attained" as two of the most fundamental aims for

---

[162] U.S. Congress, Senate, Committee on Armed Services, *Goldwater-Nichols Department of Defense Reorganization Act of 1986*, report no. 99-280, 99th Congress, 2d Session, 1986, 2-4.

establishing a command organization. In order to realize these goals the organization should provide for the following:

- Unity of effort to instill effectiveness and efficiency.

- Centralized direction to effect control and coordination of effort over the forces.

- Decentralized execution because no one commander can control the detailed actions of a large number of units or individuals.

- Common doctrines to insure mutual understanding and confidence between a commander and assigned subordinates, and among the subordinates themselves, so that timely and effective action will be taken by all concerned in the absence of specific instructions.

- Emphasis on interoperability to enhance joint warfighting [sic] capabilities through improved joint tactics, techniques, and procedures.[163]

In the recent past, these two fundamental JCS aims were not readily achievable in a given crisis situation. This being the case, the goal of the reorganization act in 1986 was to improve on the process to ensure successful military operations rather than having to settle for less-than-desired outcomes.

The chain of command within the Department of Defense, employed during both peace and war, is illustrated in Figure 4.2. This organizational flow was established in direct response to the Department of Defense Reorganization Act and portrays the line and support functions of the various principals in the combatant commands and Services with respect to the JCS and national command authorities.

Within the operational chain of command, the Unified and Specified commanders report *through* the Chairman of the Joint Chiefs of Staff to the Secretary of Defense who in turn reports to the President. The Chairman functions within the chain by transmitting orders from the national command authorities to the CINCs. Each CINC and his staff is concerned with the planning and potential employment of forces under his command during a crisis or war in his AOR. The CINC issues guidance and orders to his component commanders as authorized

---

[163] Department of Defense, Joint Chiefs of Staff, *JCS Pub. 2, Unified Action Armed Forces (UNAAF)* (Washington, DC: Department of Defense, The Joint Chiefs of Staff, December 1986), 3-2.

by his mission and the Secretary of Defense in the preparation and application of force.

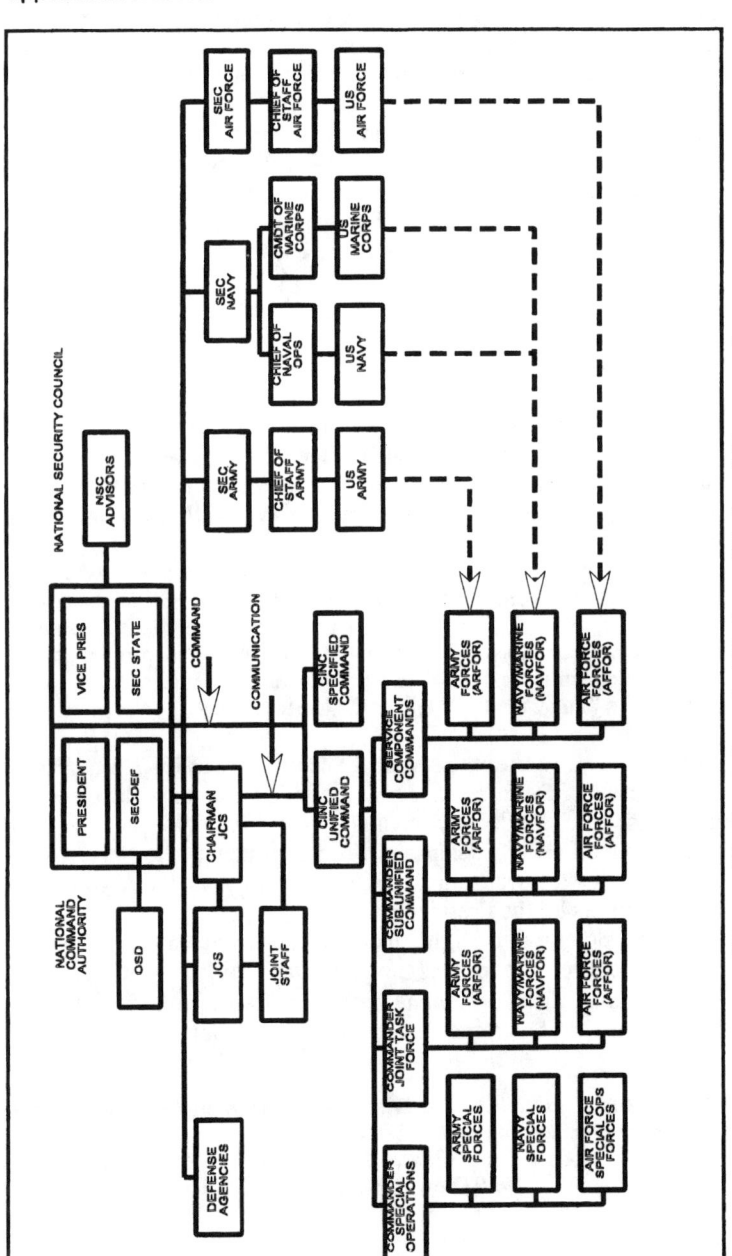

Figure 4.2 Department of Defense Chain of Command

The military Services fall under the administrative chain, which divorces them from the operational chain of command. Their mission is to organize, train, equip, and provide administrative services to the forces that would be assigned to the combatant commands during a crisis.

This overview of the chain of command is based on the way it was designed at the time of the Gulf War. With the end of the Cold War and the resulting reduction in U.S. military forces, the organization continues to evolve. (The chain of command as illustrated above and as it existed during the Gulf War will be used for the purposes of this book.)

The Political-Military Seam

The Constitution designated the President as commander of the U.S. armed forces in order to establish civilian control over the military. Civilian control was enhanced even further when, in 1789, the first U.S. Congress established the Department of War with a civilian at its head:

> There shall be an executive department to be denominated the Department of War, and that there shall be a principal officer therein, to be called the Secretary for the Department of War, who shall perform and execute such duties as shall from time to time be enjoined on, or entrusted to him by the President of the United States.[164]

Almost 200 years later the National Security Act of 1947 established the Department of Defense headed by a civilian secretary; three service departments, also headed by civilian secretaries; and the National Security Council (NSC) composed of the President, the Secretary of Defense, the service secretaries, and other defense and foreign policy officers appointed by the President.

The act, along with slight modifications enacted by follow-on legislation, created the current civilian framework of the *executive* decision-making chain of command that interfaces with the military establishment. The same act, with more substantial revisions

---

[164] *An Act to Establish an Executive Department, to be Denominated the Department of War, August 7, 1789*, 1, Stat. 7, sec. 1 (1789), quoted in Frederick C. Mosher, ed., *Basic Documents of American Public Administration, 1776-1950* (New York: Holmes & Meier Publishers, Inc., 1976), 35.

instituted primarily by the 1986 Reorganization Act, also established the command structure that directs the military decision process.

Even though this organization is vital to the American national security decision-making process, the structure laid the foundations for a cultural divergence in the way the two groups address crisis situations, with the President and the Secretary of Defense bridging the gap between the two bodies.

In many respects, this cultural difference is natural and to be expected. For example, the President and his civilian advisors often come from a more liberal persuasion than do members of the military establishment, and therefore harbor different viewpoints and values. In addition, with the advent of the all-volunteer army, current and future Presidents, along with their civilian advisors, will, more likely than not, have never been exposed to the military environment; consequently they will understand and sympathize with it less.

Richard J. Barnet profiled the social and occupational characteristics of what Kegley and Wittkopf call "policy-making elites,"[165] those people who have overseen American foreign policy since the end of World War II. Barnet wrote:

> The temporary civilian managers who came to Washington . . . the national security managers, were so alike one another in occupation, religion, style, and social status that...apart from a few Washington lawyers, Texans, and mavericks, it was possible to locate the offices of all of them within fifteen city blocks in New York, Boston, and Detroit. Most of their biographies in Who's Who read like minor variations on a single theme, wealthy parents, Ivy-League education, leading law firm or bank.[166]

On the military side of the spectrum, Samuel Huntington gave an equally excellent description of the military man in his classic book, *The Soldier and the State*. Although long, this passage captures the essence of what influences the typical military leader in the decision-making process. Huntington wrote:

---

[165] Charles W. Kegley, Jr. and Eugene R. Wittkopf, *American Foreign Policy, Pattern and Process*, 3rd ed. (New York: St. Martin's Press, Inc., 1987), 261.

[166] Richard J. Barnet, *Roots of War: The Men and Institutions Behind U.S. Foreign Policy* (Baltimore: Penguin Press, 1972), 48-49; quoted in Charles W. Kegley, Jr., and Eugene R. Wittkopf, *American Foreign Policy, Pattern and Process*, 3rd ed. (New York: St. Martin's Press, Inc., 1987), 263.

The military ethic emphasizes the permanence, irrationality, weakness, and evil in human nature. It stresses the supremacy of society over the individual and the importance of order, hierarchy, and division of function. It stresses the continuity and value of history .... It holds that war is the instrument of politics, that the military are the servants of the statesman, and that civilian control is essential to military professionalism. It exalts obedience as the highest virtue of military men. The military ethic is thus pessimistic, collectivist, historically inclined, power-oriented, nationalistic, militaristic, pacifist, and instrumentalist in its view of the military profession. It is, in brief, realistic and conservative.[167]

Cultural differences go beyond the ideology and ideals of the people involved. The advisors and staffs of both the President and top military leadership face issues within the organization that could cause discord in the workplace and adversely impact the quality of decision making. For example, on the civilian side, the career bureaucrat may feel threatened by and resentful of the political appointee. Richard Betts described the dilemma as follows: "At issue is the tradeoff between control and expertise. Imbalance on either side may have positive or negative effects, depending on the particular values and expertise involved."[168]

The same holds true within a military organization such as the Defense Department. Ronald Stupak and Thomas Hone identified the potential for tension among three groups in the department: the appointed political bureaucrat, the career civilian bureaucrat, and the "armed bureaucrat," the military professional (officer) who serves on staffs in the Pentagon and elsewhere in the Washington, D.C. area.[169] The same obstacle of control versus expertise has the potential of existing between the political appointee and the career bureaucrat and military officer in the Defense Department as it does in other federal government agencies. But there is also a high probability of tension developing between the career civilian and the military professional. According to Stupak, the catalysts include: differing perspectives at the executive level, specialist versus generalist views, decision-making

---

[167] Samuel P. Huntington, *The Soldier and the State* (New York: Vintage Books, 1959), 79.

[168] Richard K. Betts, *Soldiers, Statesmen, and Cold War Crises* (Cambridge: Harvard University Press, 1977), 41.

[169] Ronald J. Stupak and Thomas C. Hone, "National Security and Domestic Policy-Making: The Similarities and the Critical Differences," *International Journal of Public Administration* 15, no. 7 (1992): 1441-1448.

competencies, cross-structural ignorance of the other's culture, and differing worldviews.[170]

A fundamental feature of the civilian/military seam involves the factors each actor must consider when confronted with a critical decision. For example, the President must not only consider military implications of a decision (a perspective he may or may not have a thorough understanding of), but must also weigh the consequences of his decision from a political, economic, and social context. On the other hand, the primary concern of the Chairman of the JCS and his theater commander is to conduct the battle in the best way to achieve a quick victory with minimum casualties. Figure 4.3 portrays a number of the groups and factors that may have an influence on any given decision.

A final important item impacting the decision-making process concerns the personalities of the people involved. Will the President accept advice from his civilian advisors? His military advisors? How qualified are the advisors the President selected for his administration? Does the President respect his JCS Chairman and the theater commander charged with conducting military operations? Is the President a micro-manager or does he give broad direction and then allow his subordinates to carry out the task? How do the primary actors perform under stressful situations? What personal values do the actors consider important? The answers to these personality-driven questions, and many more like them, influence the decision-making process, no matter how well the system is designed.

---

[170] Ronald J. Stupak, "Military Professionals and Civilian Careerists in the Department of Defense," *Federal Manager's Quarterly* (November 1988), 19-26.

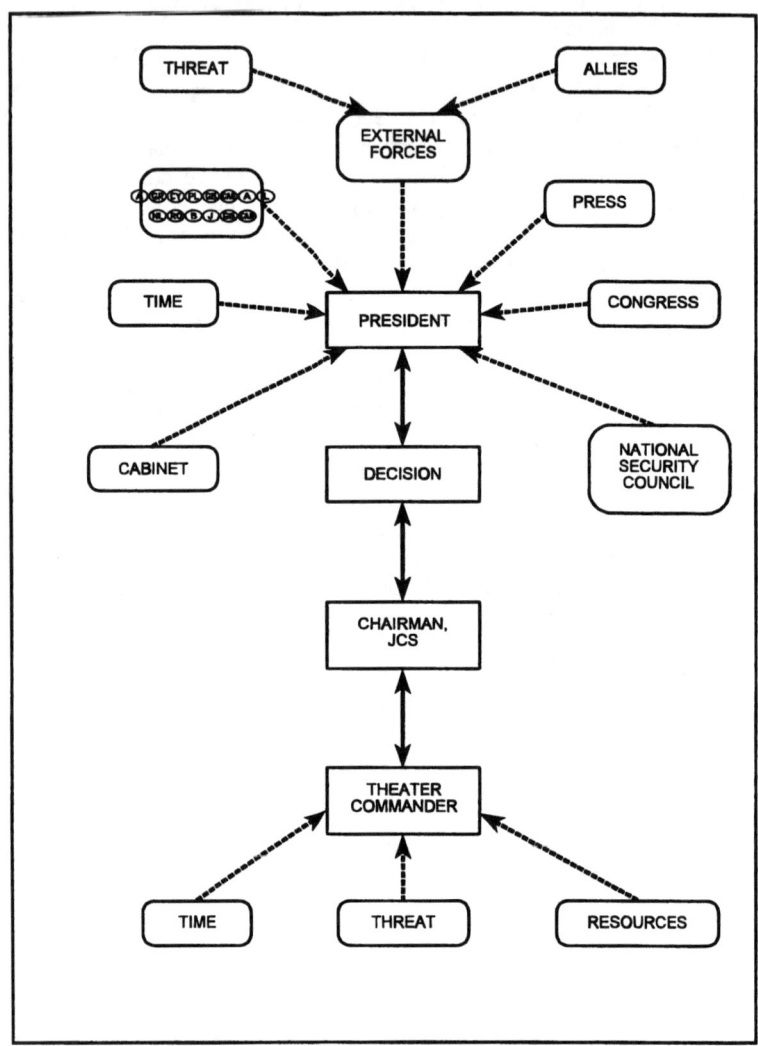

Figure 4.3 Factors Influencing Decision Making

U.S. Military Doctrine

The United States military subscribes to what best would be called the Western concept of military doctrine, which is based primarily on classical strategic theory. For 10 years now U.S. military schools have been placing greater emphasis on the value of classic strategy and the levels and principles of war. The significance placed on the fundamentals of warfare has become even more apparent with the publication of *JCS Pub 1* and *Air Force Manual 1-1*, and the revision to *FM 100-5*, all stressing the importance of a basic understanding of strategy and doctrine.

The U.S. military has historically used two specific applications of force that, while not uniquely American in origin, are synonymous with the way it chooses to fight: annihilation strategy and the application of the direct, versus the indirect, approach in attacking the enemy. Russell Weigley maintains that throughout most of U.S. history American strategists have used the concept of annihilation in the conduct of war.[171] This view is based on the theory of strategy as defined by the German military historian Hans Delbrück in the latter part of the 19th century. Delbrück, who based his concept on Clausewitz's belief in two kinds of war (total and limited), distinguished between the strategy of "annihilation" and the strategy of "exhaustion," or as Weigley defines it, "attrition." Simply stated, the strategy of annihilation seeks the destruction of the enemy's forces, while the strategy of exhaustion uses battle and maneuver as merely one of several means to attain political goals.[172]

A fine line exists between the requirement to seek the political objective and the need to properly fight the war to defeat the enemy in order to attain the political objective. The two were practically the same in World War II, so the conduct of the campaigns and the drive to achieve political objectives complemented one other. Such was not the case in Korea and Vietnam.

Several strategic thinkers, Liddell Hart being the most prolific, have extolled the virtues of the indirect application of force over the direct approach. Taking his cue from Sun Tzu, Hart stated:

> A move around the enemy's front against his rear has the aim not only of avoiding resistance on its way, but in its issue. In the

---

[171] Weigley, xxii.

[172] Gordon A. Craig, "Delbrück: The Military Historian," in *Makers of Modern Strategy, from Machiavelli to the Nuclear Age*, ed. Peter Paret (Princeton, NJ: Princeton University Press, 1986), 341.

profoundest sense, it takes the line of least resistance. The equivalent in the psychological sphere is the line of least expectation ....

In the study of the physical aspect we must never lose sight of the psychological, and only when both are combined is the strategy truly an indirect approach, calculated to dislocate the opponent's balance.[173]

Contrary to Hart's position, the American psyche demands swift execution and unqualified success when it comes to war. The direct approach to offensive action naturally follows this tendency and is the way the U.S. military conducts military operations.

The objective of total victory during World War II in both the European and Pacific theaters is a perfect illustration of this view. This strategic philosophy, in fact, caused some friction between the U.S. leadership, who wanted to attack the Germans head-on in the Mediterranean region, and the British, who preferred an indirect approach through Greece, based on Liddell Hart's strategic theory. The philosophical penchant for directness does not negate the importance of maneuver, however.

The South African National Security Policy Process

Heavily influenced by the military establishment at the time, South African national security policy played a significant role during the Angolan War, not only in the customary way state governments interact on the international political scene, but also in the way the war was conducted on a military level. To understand the significance of how political policy influenced the conduct of the Angolan War, one must have a general understanding of South African statecraft and the political/military relationship.

The South African Political Process and Decision Making

The Central Government

According to the 1983 South African Constitution, the State President is vested with both legislative and executive powers. Act no. 110, sect. 30 states, "The legislative power of the Republic is vested in the State President and the Parliament of the Republic, which, as the

---

[173] Hart, 327.

sovereign legislative authority in and over the Republic, shall have full power to make laws for peace, order and good Government of the Republic." In his legislative capacity, the State President sends bills to Parliament and signs all bills into law following approval by that body. As nominal head of state he acts on the advice of the Cabinet, which *should* limit his role in foreign affairs to largely ceremonial responsibilities. The State President's constitutional powers include appointing, accrediting, and receiving ambassadors and other diplomatic officers; entering into and ratifying treaties and agreements; and declaring war and making peace. He is also the commander in chief of the South African Defence Force and appoints ministers and deputy ministers. Elected by an electoral college consisting of members of the House of Assembly, the State President holds office for a maximum of five years.

At the time of the Angolan War, South Africa had a tricameral Parliament embodying three Houses: the House of Assembly made up of and representing "white" citizens, the House of Representatives comprised of members representing the "coloured [sic]" groups of the population, and the House of Delegates representing the "Indian" population.[174] The House of Assembly was structurally and procedurally modeled, to a large extent, on the British House of Commons. The composition of the Assembly included 165 directly elected Members of Parliament (MPs), four members nominated by the State President, and eight members chosen by the 165 directly elected MPs according to the principle of proportional representation. (Figure 4.4 illustrates the structure of the state government.)

---

[174] When the National Party took control of the government in 1948, it advanced the implementation of an apartheid policy where white and black citizens were separated physically, culturally, and economically. The classification of South African citizens, defined by race, by the Population Registration Act of 1950 identified four population groups: White, Black, Coloured, and Indian. As such, South African society was ranked in a hierarchical structure by race. Whites, who dominated the political and economic power of the nation, were located at the top. The coloureds, a mixed race created from the sexual union between whites and blacks throughout modern South African history, and the Indians, descendants of laborers brought from India beginning in the mid-1860s, formed the middle stratum of the hierarchy. Finally, the black population, consisting of members from the original African tribes, formed the bottom rung of the society.

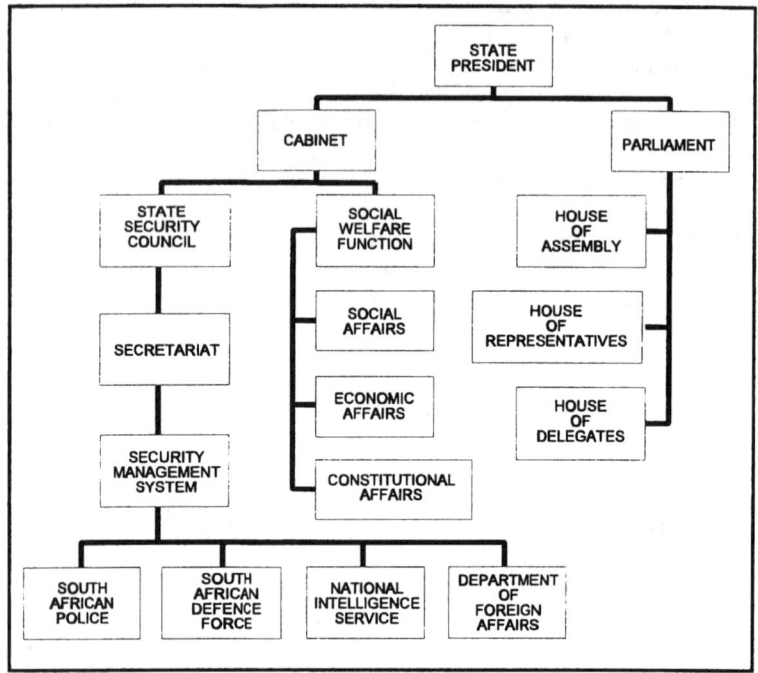

Figure 4.4 Representation of the State Government

The political process, in reality, strengthened the majority party's control over government. During the 1980s the National Party's overwhelming numerical superiority in the House of Assembly constrained the role of the chamber in foreign affairs and provided for the dominance of the State President. It also limited the ability of the House of Representatives and the House of Delegates to influence the legislative process. Parliament's primary role became then not to influence public policy but to legitimize it.

The National Party's tight control of both domestic and foreign affairs in government during the 1980s and the enhanced power held by domineering personalities in leadership positions within the party and government played a significant role in the way South Africa reacted to both external and internal events. During this period, South Africa moved in the direction of a totalitarian state, and this situation led to a national strategy and military-dominated security doctrine that embodied an aggressive foreign policy with respect to neighboring states.

The Power of the State President

In theory, under the 1983 South African Constitution, the central government was based on sound democratic principles (with the exception of the apartheid laws already on the books and the disenfranchisement of the black and colored populations). In reality, however, the government came to be dominated by the State President. Albert Venter, Associate Professor of Political Science at Rand Afrikaans University, summarized the situation as follows:

> In South Africa there is a marked tendency towards Cabinet and executive-bureaucratic dominance. The power of Parliament and the lower levels of public authority such as municipalities and provinces have systematically been usurped. Judicial checks have been devalued to ones of procedure, with the result that the central government has concentrated enormous power in its hands. Some of this is inevitable in the political environment of the late twentieth century, but the absence of democratic controls in the end makes for oligarchy and elitism. Instead of the government being the servant of the people, the people become its servant.[175]

This appropriation of power, beginning with the National Party's election victory in 1948, increasingly revolved around a smaller group of actors as the perceived threat to South Africa's stability increased in the 1970s.

The State Security Council

During this period, the principal and most important Cabinet committee in the government became the State Security Council (SSC), a group some would consider analogous to the National Security Council in the United States. The council was chaired by the State President and consisted of the following permanent members: the Ministers of Defence, Foreign Affairs, Justice, and Law and Order. By law, the council also provided a formal path for input into the executive decision-making process through the Chief of the Defence Force, Commissioner of Police, and Director General of the National Intelligence Service. The statutory function of the council, as defined

---

[175] Albert Venter, "The Central Government: Legislative, Executive, Judicial and Administrative Institutions," in *South African Government and Politics*, ed. Albert Venter (Johannesburg: Southern Book Publishers (Pty) Ltd., 1989), 94.

in the *Security Intelligence and State Security Council Act* of 1972, was to advise the state government in the formulation and implementation "of national policy and strategy in relation to the security of the Republic."[176] In other words, in theory, decisions should have been referred to the Cabinet for ratification. In reality, the council took on an increasingly more important and expanded role, chiefly because of problems resulting from decisions made during the early days of the Angolan War and the struggle for power by egocentric personalities dominating the government since the late 1970s.

The SADF and the State Government

With South Africa divorced from the community of nations due to its policy on apartheid, the South African Defence Force became heavily involved in policy making within the national government. As a result, South African national strategy and foreign policy took on a more aggressive stance in the 1970s and 1980s, fueled by a long-perceived threat of communism.

Cuban involvement in Angola and Soviet influence within the African National Congress (ANC) organization forced the government to act. Under Minister of Defence P.W. Botha and Chief of the SADF General Magnus Malan, the government instituted a "total strategy" policy in 1977 that became a blueprint to counter the "total onslaught" concept thought to be the design used by Marxist-led countries and organizations to defeat South Africa.[177] This total strategy policy came directly from strategic theory advocated by André Beaufre, in his concept of "indirect strategy."[178] It enabled the military to become a pivotal player in the decision-making process on both foreign and domestic issues.

According to Malan, the plan's architect, every activity of the state and the private sector would become a function of the strategy, which would require the reallocation of social, political, economic, and manpower resources in order to fight the "total war." This strategy resulted in a more extensive South African military effort in Angola.

Many students of the South African political scene believed that the military was at the center of decision making in the government in the

---

[176] Deon Geldenhuys, *The Diplomacy of Isolation, South African Foreign Policy Making* (Johannesburg: Macmillan South Africa (Publishers) (Pty) Ltd., 1984), 92.

[177] T.R.H. Davenport, *South Africa, A Modern History*, 3d ed. (Toronto: University of Toronto Press, 1987), 438.

[178] Brigadier Willem S. Van der Waals (SADF, Ret.), interview by author, 9 July 1994, Pretoria, South Africa.

mid-1980s. P.H. Frankel, for example, wrote, "Militarisation is always measured by the appearance of soldiers as public decision-makers, and the growing influence of the South African military is finally, and perhaps most importantly reflected in the penetration of top government institutions by Defence Force personnel, on either a formal or informal basis."[179] Several years later, Vale asserted in the *Star* newspaper that the military constituted an extra-parliamentary government that actually ruled South Africa.[180]

A military-controlled government can be overt (i.e., a military dictatorship) or it may assume a more subtle role in a more liberal form of government. In the latter case, "the armed forces do not occupy the front line in the political sense. They do not govern directly, but exercise rather tight control over the formal holders of power."[181] Critics and adversaries of the state government contended that this was the case in South Africa. They pointed to the SSC where the SADF controlled the operating strategies and resources of government. They maintained that the SSC, whose secretariat was largely manned by the SADF, replaced the State President's Cabinet as the most influential decision-making body in the bureaucracy. In effect, the SSC, they argued, worked directly for and influenced a limited number of government officials, including the State President, Minister of Defence, and Minister of Foreign Affairs. This then was the seat of power and the de facto decision-making body during the Angolan War. For all intents and purposes, the critics concluded, the normal democratic process had been relegated to a small number of individuals in the government who were not subject to the normal checks and balances expected in a democracy.

The South African Defence Force and Its Doctrine

In 1957 the Defence Act was passed by Parliament and signed into law, becoming the statutory authority for the South African Defence Force. The Act determined the legal position of the SADF, sanctioned the existence of its permanent and citizen forces, created a new military

---

[179] P.H. Frankel, *Pretoria's Praetorians: Civil-Military Relations in South Africa* (Cambridge: Cambridge University Press, 1984), 103.

[180] *Star* (Johannesburg), 12 February 1988.

[181] M. Lowy and E. Stader, "The Militarization of the State in Latin America," *Latin American Perspective* 12, no. 4 (1985): 9, quoted in Jacklyn Cock and Laurie Nathan, eds., *War and Society: The Militarisation of South Africa* (Claremont, S.A.: David Philip, Publisher (Pty.) Ltd., 1989), 8.

code of discipline, and instituted the commando structure.[182] Several sections of the Act applied to the conduct of national security policy and are highly germane to this discussion:

- Section 3 of the Act became extremely important in recent South African history as it allowed the SADF to be employed "in the prevention or suppression of internal disorder in the Republic."[183] This precipitated a high-profile use of military force for the preservation of internal stability
- Section 1 of the Act gave the State President wide latitude to use the Defence Force to defend the Republic ". . . in time of war or in connection with the discharge of the obligations of the Republic arising from agreement or for the prevention or suppression of any armed conflict outside the Republic which, in the opinion of the State President, is or may be a threat to the security of the Republic."[184]
- Section 118 of the Act gave the State President far-reaching powers over communications and the media in South Africa. The Act prohibited persons from publishing information on the composition, movements, and disposition of the SADF. It also presumed that "any information relating to the defence of the Republic is secret or confidential."[185]

Composition of the Defence Force

The SADF was divided into two distinct components, a "full-time" permanent force and a "part-time" reserve force. The full-time component, equivalent to the regular armed forces of the United States, consisted primarily of career soldiers and national servicemen who had been conscripted for a two-year tour of duty. The Part-time Force, roughly comparable to the U.S. Reserves and National Guard (with some qualifications), was composed of a citizen force and commandos. Figure 4.5 compares the make-up of the two forces. The relatively small number of permanent force members would normally exacerbate

---

[182] Kathy Satchwell, "The Power to Defend: An Analysis of Various Aspects of the Defence Act," in *War and Society: The Militarisation of South Africa*, ed. Jacklyn Cock and Laurie Nathan (Claremont, S.A.: David Philip, Publisher (Pty) Ltd, 1986), 40.

[183] Parliament, *Defence Act No. 44 of 1957, Statutes of the Republic of South Africa - Defence*, 13, sec. 3.(2)(a)(iii)(1957).

[184] Ibid., sec. 1.

[185] Ibid., sec. 118(5)(a).

SADF, in practice it typically fulfilled 87 percent of the service requirements. This imbalance resulted from high deferment rates, administration practices, and budget restraints on the Part-time Forces. In a crisis situation, this resource/commitment asymmetry would place a heavy burden on the National Servicemen component of the Full-time Force, a situation that impacted the SADF's ability to fight during the Angolan War.

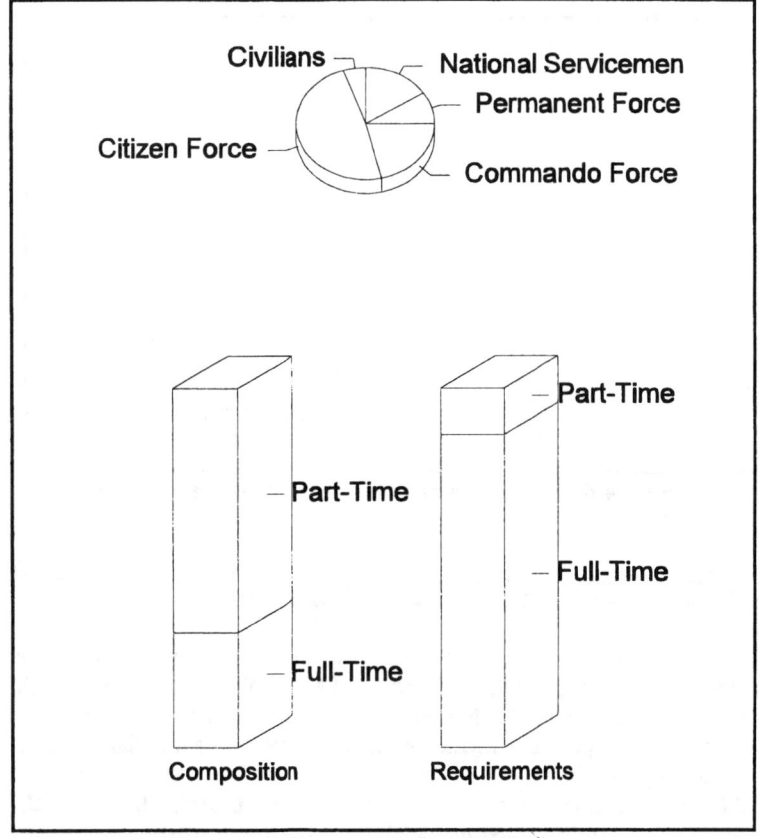

Figure 4.5 Composition of SADF Forces

SOURCE: Republic of South Africa, Minister of Defence, *White Paper on Defence and Armaments Supply — 1986* (Cape Town, S.A.: Naval Printing Press, 1986), 4.

The SADF was mainly dependent on the white male as a source of manpower for both the Full-time and Part-time Forces. Figure 4.6 illustrates the racial make-up of the Full-time Force in 1986. It was this heavy reliance on white males that also caused operational problems when the South African government escalated its involvement in Angola.

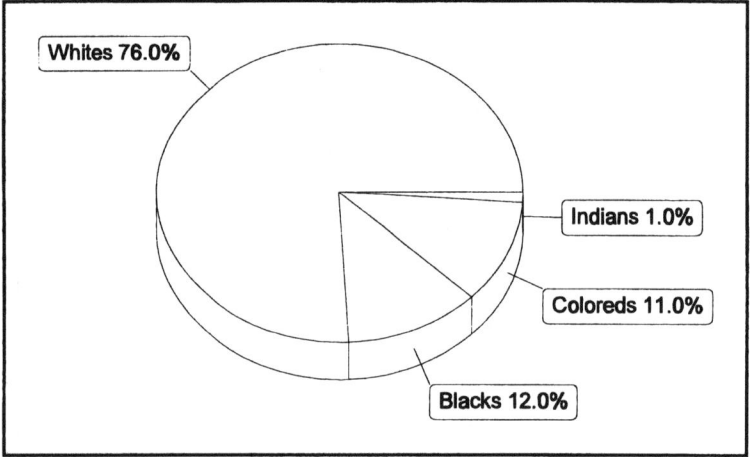

Figure 4.6 Racial Composition of the Full-Time Force

SOURCE: Republic of South Africa, Minister of Defence, *White Paper on Defence and Armaments Supply—1986* (Cape Town, S.A.: Naval Printing Press, 1986), 17.

The permanent force of the SADF was (and still is) made up of four branches, the Army, Navy, Air Force, and Medical Services. The chiefs of each service reported to the Chief of the Defence Force who, in turn, was directly responsible to the Minister of Defence, a civilian cabinet position. (See Figure 4.7.) In an extreme national emergency, the SADF was capable of mobilizing from 400,000 to 500,000 personnel. This number could not be matched by any country in southern Africa without foreign assistance. But just as important, the SADF consisted of a professional, well-trained, and well-equipped corps of soldiers, sailors, and airmen. Their equipment, although fairly old, was well maintained. The Armaments Corporation of South Africa (Armscor), an organization falling under the Ministry of Defence, developed and produced weapons for the SADF that in many cases are still world-class. In short, there was no indigenous army south of the Sahara Desert that was capable of posing a significant military threat to South Africa during the Angolan crisis.

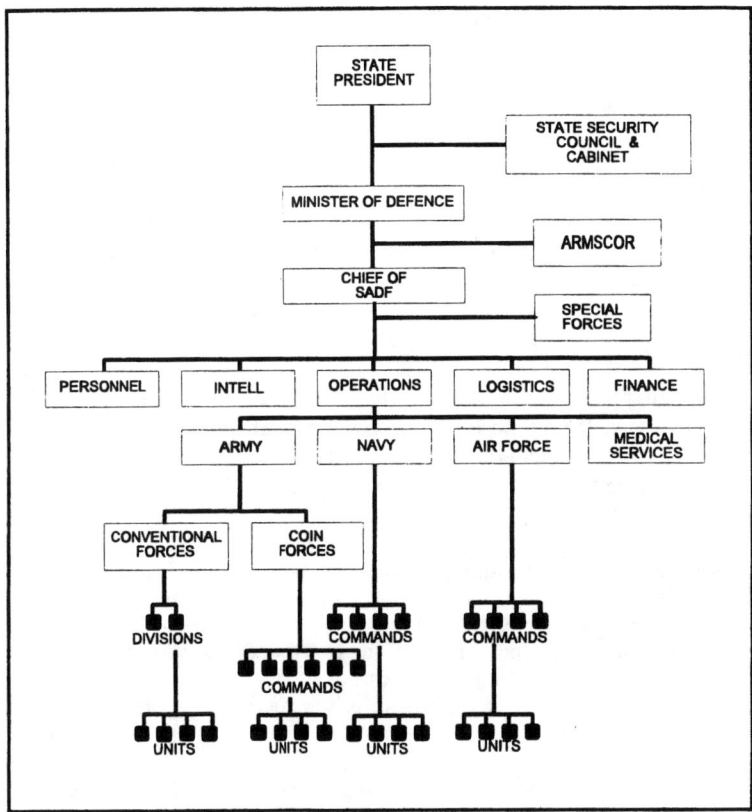

Figure 4.7 SADF Organizational Structure

SADF Doctrine

While a member of the British Commonwealth, South Africa's military doctrine and national security strategy were dominated by

British doctrine.[186] But when it withdrew from the Commonwealth in 1961, a vacuum was created in doctrinal development as the SADF became isolated from other militaries around the world. Strategic thought became stagnant and remained that way until the late 1960s and early 1970s when military thinkers discovered General André Beaufre and his indirect strategy of using economic and political means as well as military force.[187]

The majority of current SADF doctrine is either classified or restricted, so minimal official information is available. What is known indicates that the SADF employs the Western doctrinal model-- basically the same as that used by the United States and NATO, but with some modification for the African bush and local terrain. Also known is that the SADF Military Academy and professional staff colleges teach classic theory based on the works of Sun Tzu, Clausewitz, Jomini, and Mahan. These works are augmented with strategic concepts advanced by more recent military theorists such as Douhet, Hart, Fuller, and Beaufre in particular. Along with classic strategy, South African officers are very familiar with the "levels of war," the importance of command and control, and the value of technology.

The principles of war advanced by the SADF (exhibited in Table 3.1) are derived from both U.S. and British doctrine. They are rooted in military theory; nonetheless, the army takes a very pragmatic view on their applicability and limitations. South African conventional doctrine states:

> Mere knowledge and understanding of the principles of war will not provide the solution to every problem of war. The human elements, courage, morale, discipline, and leadership, have a direct bearing on the outcome of any operation and are so vital to success that they deserve constant attention. In the final analysis, sound judgment and common sense are of vital importance to the successful application of the principles of war.[188]

Conventional war based on the NATO concept would not work in many areas of southern Africa because of dense bush, rugged terrain, and inhospitable weather. The SADF therefore modified the Western doctrinal concept to one based on British counterinsurgency (COIN) operations in Malaya. By relying on this strategy, South Africa was

---

[186] Van der Waals, interview by author, 9 July 1994.
[187] Beaufre, 44.
[188] Republic of South Africa, South African Army, *Conventional Land Battle*. (Pretoria: South African Army), 1-44.

able to frustrate the expansion of communism into southern Africa for 16 years, from 1973 until a peace treaty was signed in 1989. According to Brigadier General George Kruys, an SADF warrior and strategist, South Africa succeeded in the Angolan War by emphasizing high levels of training and employing combat troops who had a lot of initiative.[189] In a summary of the strategy used in the war, he concluded, "The SA Government kept communism at bay using a strategy of low intensity military operations, at great distance, using high technology and using time brilliantly. It did not use lives and money indiscriminately causing a war exhaustion as experienced in many previous conflicts overseas."[190]

South African National Security Policy, Total Strategy

A number of significant operations involving a host of perceived threats to the South African government took place over several fronts in the mid-1970s and 1980s, culminating in the intensification of war in Angola during 1987/88. The Soviet Union, for instance, attempted to spread its influence throughout southern Africa with the declared intent of controlling the mineral wealth of the region. It also supplied the Marxist-led government in Angola with significant logistical support to resist an insurgency backed by the West, and was heavily involved with inciting revolution in the "front-line states" on South Africa's northern border. Meanwhile, Cuba placed a sizable army in Angola to support the government against Jonas Savimbi (a sometime ally of South Africa) and his guerrilla forces.

The Angolan government provided a safe haven for insurgent forces intent on forcing South Africa out of South West Africa (Namibia). The government also provided sanctuary and training for members of the African National Congress, an organization intent on fomenting revolution against the South African government and taking political control of the country.

The South African Department of Defence *White Paper on Defence and Armaments Supply 1984* described the threat to South Africa as follows:

One of the major considerations of Soviet strategy with regard to Southern Africa is the control of the subcontinent's riches in strategic minerals and the denial of these to the West. In this regard President Brezhnev stated plainly in 1977:

---

[189] George Kruys, "Fallacies Are a Matter of Twisted Facts," *Paratus* (April 1993), 19.
[190] Ibid., 22.

"Our aim is to gain control of the two great treasure houses on which the West depends, the energy treasure house of the Persian Gulf and the mineral treasure house of Central and Southern Africa." [It is interesting to note that the SADF has never given a reference as to where and under what circumstances Brezhnev made this quote.]

Indirect action in the form of a revolutionary onslaught serves to establish Soviet influence in Southern Africa. The South African Communist Party (SACP) and the African National Congress (ANC), which, for all practical purposes, has been integrated with the SACP and acts as its military wing, are the major elements of the Soviet plan to obtain control of the RSA. SWAPO plays a similar role in SWA in the achievement of Soviet objectives in that region. Several worldwide and regional organizations, of which the United Nations (UN) and the Organization for African Unity (OAU) are the most important, also lend themselves to furthering USSR objectives in Southern Africa by joining in the propaganda onslaught against the RSA.[191]

Beginning in the late 1970s, the Pretoria government's response to these events was to develop a national security policy identified as a "total strategy" to counter this "total onslaught" against South Africa from a myriad of directions. Critics of the government maintained, however, that this strategy "supplied the basis for legitimising [sic] an increasing military involvement in all spheres of national, regional and local government decision-making. In this sense total strategy was the launchpad for the militarisation [sic] of South African society."[192]

The Actors

Personalities played a significant role in the South African government's decision making during the previous two decades. Individuals having the greatest impact at the national level were as follows:
- Prime Minister
  - B.J. Vorster[193]   1966 - 1978
  - Pieter W. Botha   1978 - 1984

- State President[194]

---

[191] South Africa Department of Defence, *White Paper on Defence and Armaments Supply 1984* (Cape Town, S.A.: Naval Printing Press, 1984), 1.
[192] Cock and Nathan, 2.
[193] As a previous Minister of Police, B.J. Vorster relied heavily on the South African Police (SAP) as his power base.

- Pieter W. Botha   1984 - 1989

- Minister of Foreign Affairs
  - R.F. (Pik) Botha   1976 - 1994

- Minister of Defence
  - Pieter W. Botha   1966 - 1980
  - Magnus Malan   1980 - 1990

- Chief of the Defence Force
  - General Magnus Malan   1976 - 1980
  - General Constand L. Viljoen   1980 - 1985
  - General J.J. "Jannie" Geldenhuys   1985 - 1990

When P.W. Botha was elected Prime Minister in 1978, he set about reorganizing the central government by giving the security establishment, especially the SADF which served as his power base, a more influential role in decision making. He elevated the Chief of the Defence Force, General Magnus Malan, to the position of Minister of Defence. In turn, Malan in 1980 identified the SSC and its substructures as the key organization for implementing the total strategy concept.

Power in the government was consolidated in the executive branch and the security establishment. The SADF, the intelligence community, the intellectual community (from primarily Afrikaans universities), Armscor, and the South African Police constituted the "primary coordinating and binding force within the state."[195] Although the security forces became a dominant force in the government, civilian control was maintained through the Prime Minister/State President. Although critics insisted that "while the government kept alive the 'sham of democracy,' the security establishment took overall control of strategy and decision-making in the state."[196]

---

[194] The new South African constitution adopted in 1983 changed the function of head of state from a non-executive "Prime Minister" to an executive "State President." The prime minister function was absorbed by the new State President.

[195] Mark Swilling and Mark Phillips, "State Power in the 1980s: From 'Total Strategy' to 'Counter-revolutionary Warfare,'" in *War and Society, The Militarisation of South Africa*, ed. Jacklyn Cock and Laurie Nathan (Claremont, S.A.: David Philip, Publisher (Pty.) Ltd., 1989), 137.

[196] Ibid., 138.

Chapter 5

The Gulf War Case

The Persian Gulf War can be divided into three discrete operations: the buildup, the war itself, and the consequences following war termination. This chapter will describe the events of those operations relevant to the strategic and operational strategy of conducting the war effort. But first, a brief introduction to events preceding the crisis needs to be presented to set the stage for American involvement in the region.

Prelude to Invasion

In the pre-dawn hours of 2 August 1990, three Iraqi Republican Guard divisions invaded the Kingdom of Kuwait and subsequently gained control of the country in less than 36 hours. The invasion set in motion a U.S. political and military response unparalleled in post-World War II international relations. The scope and success of the ensuing military buildup and diplomatic initiatives to forge a coalition of forces in opposition to Iraqi aggression surprised even those in Washington tasked with developing a response to the invasion.

In retrospect, three unrelated events on the international scene conspired to influence the crisis, from Iraq's decision to invade Kuwait to the United States' ability to counter the aggression. Those events were:

- Iraq's designs on the country of Kuwait for historic and economic purposes
- A perceived threat to United States vital interests
- The end of the Cold War.

Iraq Covets Kuwait

Iraq claimed jurisdiction over Kuwait based on the fact that they both belonged to the Ottoman Empire before its collapse during World War I. When Kuwait received its independence in 1961, Iraq's ruler at the time, General Abdul Karim Qassin, declared Kuwait to be Iraqi territory, and only military support from Great Britain prevented Iraq from taking possession of the country. Even though Kuwait subsequently became a member of the United Nations and the Arab League, Iraq never relinquished its claim to the territory.

Iraq's eight-year war with Iran from 1980 to 1988 provided an economic catalyst for the invasion of Kuwait. Bruce Watson suggested five reasons motivating the decision to invade:

- Iraq could not repay the $80 billion that it had borrowed to finance the war with Iran; $65 billion of that total was owed to Kuwait.
- Access to Kuwait's wealth would solve Iraq's debt problems.
- Kuwait was drilling in the Rumaila oil field, which was in a disputed border area between the two countries.
- Kuwaiti overproduction of oil was exacerbating an already significant oil glut that was depressing the price of oil on the world market.
- When Saddam Hussein proposed peace talks, the Emir of Kuwait rejected face-to-face talks in favor of mediation by the Arab League.[197]

U.S. Vital Interests

In 1943 President Franklin D. Roosevelt declared, "The defense of Saudi Arabia is vital to the defense of the United States."[198] From that time forward, the Middle East became an area of interest to the United States, with presidential doctrine from Truman to Carter evolving in response to the changing threat of the Cold War and the increased importance of the Persian Gulf area as a result of the West's increased dependency on oil.

President Carter, in particular, emphasized the significance of the Middle East to U.S. foreign policy with the establishment of the Rapid

---

[197] Bruce W. Watson and Bruce W. Watson, Jr., "The Iraqi Invasion of Kuwait," in *Military Lessons of the Gulf War*, ed. Bruce W. Watson, 2d rev. ed. (London: Greenhill Books, 1993), 17.

[198] Joe Stork and Martha Wenger, "From Rapid Deployment to Massive Deployment," in *The Gulf War Reader*, ed. Micah L. Sifry and Christopher Cerf (New York: Random House, Inc., 1991), 34.

Deployment Force (RDF). In his 1980 State of the Union address, he proclaimed that "an attempt by any outside force to gain control of the Persian Gulf region will be regarded as an assault on the vital interests of the United States, and such an assault will be repelled by any means necessary, including military force."[199] Fortunately there was no challenge to President Carter's pronouncement, because at that time the RDF was little more than a paper concept. It would take another eight years of increased military spending by the Reagan Administration to field a force sufficiently powerful enough to prevail in a conventional war halfway around the world.

The United States became an active player in the Middle East during the Reagan years. In 1983 a new unified command, Central Command (CENTCOM), was established whose area of responsibility was the Middle East, including the area around the Persian Gulf. A force of 300,000 to 350,000 troops was allocated to CINCCENT should intervention be required. Seventeen ships were stationed near Diego Garcia to provide pre-positioned supplies for any contingency.[200] Billions of dollars were spent by both the United States and Saudi Arabia to upgrade airfields, ports, air defense networks, weapon systems, and other support facilities over the course of the decade. Additional billions were spent in other Middle Eastern countries to provide the facilities required for the direct interjection of U.S. forces from the United States should it become necessary.

Finally, in 1987, the United States became militarily involved in Persian Gulf affairs by reflagging Kuwaiti oil tankers to protect them while they transited the gulf. As a result, a U.S. naval deployment in and around the Gulf grew to nearly 50 ships.

Three years later, in the 1990 edition of the *National Security Strategy of the United States*, President Bush clearly identified the Middle East as an area of interest for the United States. According to this national document, U.S. strategy in the Middle East when Iraq attacked Kuwait in the summer of 1990 was based on the following: "The free world's reliance on energy supplies from this pivotal region and our strong ties with many of the region's countries continue to constitute important interests of the United States."[201] The document went on to clarify this position as follows: "Secure supplies of energy are essential to our prosperity and security. The concentration of 65 percent of the world's known oil reserves in the Persian Gulf means we

---

[199] *New York Times*, 24 January 1980.
[200] Ibid., 37.
[201] White House, *National Security Strategy of the United States* (Washington, DC: The White House, March 1990), 13.

must continue to ensure reliable access to competitively priced oil and a prompt, adequate response to any major oil supply disruption."[202]

The Middle East was not specifically identified as a *vital* interest to U.S. security as part of this national security strategy. The closest the document came to acknowledging this point was identifying an *enduring broad national interest* as being "a healthy and growing U.S. economy to ensure opportunity for individual prosperity and a resource base for national endeavors at home and abroad."[203] To support this enduring national interest the document declared, "National security and economic strength are indivisible. We seek to . . . ensure access to foreign markets, energy, mineral resources, the oceans, and space."[204]

End of the Cold War

The demise of the Soviet Union and ensuing end to the Cold War could not have come at a more opportune time for the United States and its allies. In fact, the ability of the United States to counter the Iraqi invasion would have been significantly different had the East-West struggle remained the same as during the previous 45 years. In that case, the Gulf War could have easily degenerated into another Korea or Vietnam, saddled with the political baggage associated with superpower confrontation.

But instead, the international scene had changed significantly, allowing the United States to counter the crisis more freely than had been possible since the Second World War. In fact, Russia, with few exceptions, actually supported the United States and Coalition countries against Iraq, its former ally. In the same vein, Russia did not veto any of the Security Council resolutions in the United Nations that condemned the aggression and instituted economic embargoes on commerce bound to and from Iraq.

In addition, because of easing East-West tensions, the United States was able to deploy significant amounts of military equipment from Europe to support the war effort without the fear of Eastern Bloc countries taking advantage of the situation. And since the two superpowers were not ideologically opposed during the war, there was little fear of the war escalating into a nuclear confrontation. This made it much easier to form a coalition of nations of diverse political, religious, and social backgrounds to oppose the Iraqi invasion.

---

[202]Ibid., 22.
[203]Ibid., 2.
[204]Ibid., 2.

Desert Shield

The United States Responds
Following Iraq's successful invasion of Kuwait in August, Iraqi troops immediately began amassing along the Kuwait-Saudi Arabian border. This action compelled the civilian and military leadership in Washington to anticipate an Iraqi invasion of Saudi Arabia and the need for U.S. intervention to defend that country. The events during the days immediately following the invasion were conducted with these suppositions in mind:

- President Bush began an immediate campaign to enlist allied and United Nations support in condemning the aggression and laying the foundation for a coalition of nations to oppose the invasion.
- On 2 August, the United Nations Security Council (UNSC) adopted Resolution 660, condemning the invasion and calling for the immediate and unconditional withdrawal of all Iraqi troops from Kuwait.
- CINCCENT, General H. Norman Schwarzkopf, and his staff began updating operations plan (OPLAN) 1002-90, an on-the-shelf contingency plan for the insertion of U.S. forces into the Middle East.
- After briefing the President on possible military options, General Schwarzkopf began preparing forces for deployment to Southwest Asia.
- On 3 August, President Bush warned Iraq not to invade Saudi Arabia, and the United States offered to defend Saudi Arabia against an Iraqi attack.
- On 6 August, Secretary of Defense Dick Cheney and General Schwarzkopf briefed King Fahd of Saudi Arabia on the military situation in Kuwait, emphasizing the size of the Iraqi force in that country and the defensive options contained in OPLAN 1002-90 for the deployment of American troops to Saudi Arabia. The King agreed to invite the United States into his country; when notified of this, President Bush approved the issue of deployment orders and the implementation of OPLAN 1002-90.

On 8 August, President Bush announced to the nation that the United States was deploying American troops to Saudi Arabia to help defend that country from a possible Iraqi attack. In his address he began laying the foundation on which he would gain public support for the military operation when he proclaimed, "Let me be clear; the sovereign independence of Saudi Arabia is of vital interest to the United States."[205] He also presented four objectives to guide U.S. policy in response to the crisis:

- Immediate, unconditional, and complete withdrawal of all Iraqi forces from Kuwait
- Restoration of Kuwait's legitimate government
- Security and stability of Saudi Arabia and the Persian Gulf
- Safety and protection of the lives of American citizens abroad.[206]

These policy guidelines governed the conduct of operations throughout both the defensive and offensive phases of the crisis from both a political and a military perspective.

The Military Buildup

Early in the crisis, CINCCENT established military objectives for Operation Desert Shield that provided guidance for planners to develop options for further military actions. These objectives were to:

- Develop a defensive capability in the gulf region to deter Saddam Hussein from further attacks.
- Defend Saudi Arabia effectively if deterrence failed.
- Build a militarily effective Coalition and integrate Coalition forces into operational plans.
- Enforce the economic sanctions prescribed by UN Security Council Resolutions 661 and 665.[207]

A Concept of Operations was developed from these objectives that initially called for an "area defense" of Saudi Arabia that would trade space for time and use air power to stem the flow of an Iraqi attack. This concept would remain in effect until sufficient Coalition forces became available to establish a more formidable defensive line at the Kuwaiti-Saudi border.

In terms of time and distance, the military buildup in Southwest Asia was the largest and most extensive projection of air power in history. According to Lieutenant General Charles Horner, commander of U.S. and Coalition air forces during the war, more tonnage was

---

[205] *New York Times*, 9 August 1990, A15.
[206] Department of Defense, *Conduct of the Persian Gulf War: Final Report to Congress* (Washington, DC: Department of Defense, April 1992), 22.
[207] Ibid., 40.

moved by air in the first six weeks of the buildup than was moved during the entire 65 weeks of the Berlin Airlift.[208]

The initial military objective of the U.S. campaign was to deploy sufficient troops to Southwest Asia to deter an Iraqi attack on Saudi Arabia, and barring that, to defend the kingdom and defeat further aggression. The initial few weeks were crucial as the first American combat troops (members of the 82nd and 101st Airborne Divisions and elements of the 1st Marine Expeditionary Force [MEF]) began arriving in-country on 8 August. They were provided air cover by Saudi F-15Cs, the USS *Independence* carrier air wing, and the initial cadre of F-15Cs of the First Tactical Air Wing that arrived in-country on the same day.

On 18 August, President Bush authorized the first ever activation of the Civil Reserve Air Fleet (CRAF) in its 38-year history, increasing strategic airlift capability with 95 passenger and 63 cargo aircraft.[209] He followed with an executive order on 22 August that authorized the call-up of 200,000 reservists for a period of up to 180 days, the largest mobilization of reserve personnel since the Korean War. By the end of August, the United States and Great Britain had the following forces supporting the Saudi defensive operation:

- Two fighter squadrons from the Royal Air Force with accompanying tanker and maritime patrol aircraft
- Fourteen U.S. Air Force and Marine Corps tactical fighter squadrons
- Three carrier battle groups
- One B-52 squadron
- Seven Army and Marine Corps brigades with attack helicopters
- One Patriot air defense system.

Those first few weeks in August and September provided Saddam Hussein with his only chance to successfully invade Saudi Arabia. The lightly armed members of the 82nd, the 101st, and the 1st MEF with minimal air support could not have withstood a concerted Iraqi advance supported by the large numbers of heavy tanks assembled along the Saudi border; but the defensive capabilities of the Coalition improved every day as more men, equipment, and aircraft came on-line. The serious situation resulting from disproportionate combat power was finally redressed at the end of September by the arrival of the 24th

---

[208] Lt. General Charles A. Horner, "The Air Campaign," *Military Review* LXXI, no. 9 (September 1991): 19.

[209] Rod Alonso, "The Air War," in *Military Lessons of the Gulf War*, ed. Bruce W. Watson, 2d rev. ed. (London: Greenhill Books, 1993), 61.

Mechanized Infantry Division with its 200 M1A1 Abrams tanks and Bradley infantry vehicles. From that point on the story of the buildup was basically one of numbers illustrating the speed and magnitude of reinforcement efforts. The initial deployment was essentially completed by 7 November.

Forging the Coalition

The military buildup in Southwest Asia was a logistics marvel in its own right; but the cohesion displayed by the coalition of countries formed to oppose Iraq was an historic first. The Administration team of President Bush, Secretary of State James Baker, and National Security Advisor Brent Scowcroft performed a near miracle in bringing together a group of nations with such diverse cultural, religious, ethnic, and ideological backgrounds.

It was Secretary Baker who recommended utilizing the speculative, yet highly successful, option of employing the United Nations as a vehicle for engineering diplomatic action against Iraq.[210] Just as important, the United States persuaded the Soviet Union to remain, at a minimum, diplomatically neutral and not to become involved militarily. In the end, 36 nations provided military support to the Coalition and others provided equipment or economic assistance.[211]

Throughout the crisis the United Nations performed laudably in its role as a world forum responsible for responding to critical political imperatives. Without a significant threat from a Soviet veto, the UN was able to pass 12 resolutions between 2 August and 29 November that, among other considerations, condemned Iraq for the invasion, imposed an embargo on all trade with Iraq, authorized a naval blockade, and, of most importance (and surprise), approved use of all means necessary to drive Iraq from Kuwait after 15 January 1991. Active UN involvement gave legitimacy to the Coalition, and, along with strong U.S. leadership, helped form and hold the alliance together.

The Chain of Command

As the Coalition grew, a dual chain of command evolved because of the varied political, military, and cultural differences among the participating countries. Generally speaking, the Islamic forces were placed under the command of Saudi Lt. General (Prince) Khalid bin

---

[210] U.S. News & World Report, *Triumph Without Victory, The History of the Persian Gulf War* (New York: Times Books, 1993), 104.
[211] U.S. Department of Defense, *Conduct of the Persian Gulf Conflict: An Interim Report to Congress* (Washington, DC: Department of Defense, July 1991), I-3.

Sultan bin Abdul-Aziz while American and non-Islamic members of the Coalition were commanded by CINCCENT, as illustrated in Figure 5.1. A Coalition Coordination Communications and Integration Center (C^3IC) was established to plan and coordinate the efforts of the U.S., British, and French forces with those under the command of the Saudi general.

A Change in Mission

Although plans for offensive action against Hussein were being developed at CENTCOM Headquarters throughout the early stages of the crisis, primary emphasis was on defensive action during those first few months as Coalition forces arrived in the theater. By the time the initial deployments were completed in early November, however, it was becoming more obvious that a military option would likely be required to force Hussein out of Kuwait. At this time, the Coalition adopted an offensive orientation, with the objective of designing a ground campaign that was "short, sharp, and decisive, with minimum casualties."[212]

While the Coalition increased the size of its force in and around Saudi Arabia, Hussein was doing the same in Kuwait. Any offensive action initiated to drive the Iraqi army from Kuwait would require more men and equipment than originally planned for in the opening days of the crisis. General Schwarzkopf was therefore authorized an additional force of 200,000 American troops, most of whom would be deployed from Europe.

The rapid buildup and sustainment of an immense American military force 8,000 miles from the continental United States would be

---

[212] Peter Tsouras and Elmo C. Wright, Jr., "The Ground War," in *Military Lessons of the Gulf War*, ed. Bruce W. Watson, 2d rev. ed. (London: Greenhill Books, 1993), 89.

108   *War as an Instrument of Policy*

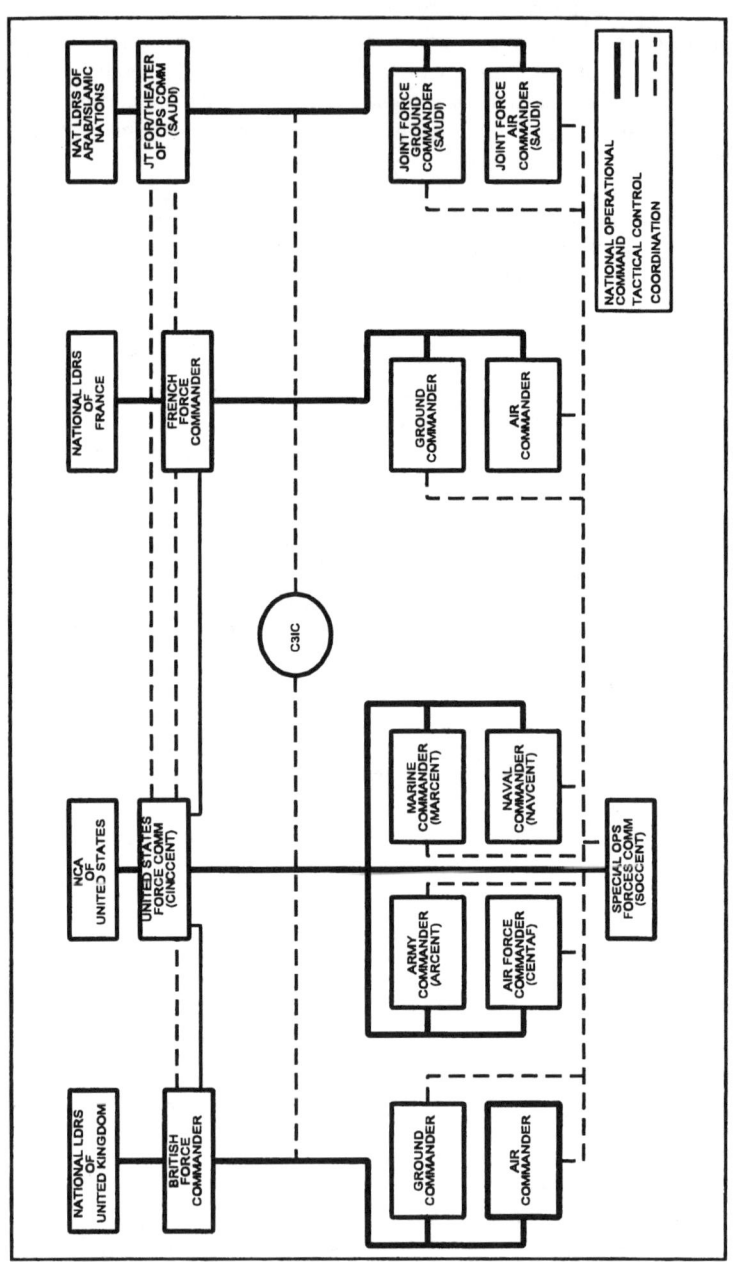

Figure 5.1 The Coalition Dual Chain of Command

one of the main legacies of Desert Shield. The six-month buildup during the operation placed a tremendous burden on the military transportation system. During that period, sea lift brought 95 percent of the equipment and supplies to Southwest Asia (by the war's end, almost three million tons of cargo), while strategic airlift carried 99 percent (over 500,000) of the personnel. Tables 5.1 and 5.2 depict the order of battle for both the Coalition and Iraqi forces when military action was initiated during the night of 17 January 1991.

The second legacy of Desert Shield was the coming together of a coalition of nations to resist naked aggression by Iraq. This alliance was even more notable since several of the Coalition nations were historically allied with Iraq and possessed the same culture, religion, and ideology. (Under extreme pressure from the United States, the United Nations performed in a socially responsible manner, an obligation that the organization had rarely adequately dealt with in the past.)

Air Forces	1,796	Combat aircraft[213]
Ground Forces	540,000	Troops
	1,200	Tanks
	2,200	APCs
	1,700	Helicopters[214]
Naval Forces	100+	Ships
	6	Carrier battle groups
	2	Battleship surface action groups
	13	Submarines
	4	Amphibious formations[215]

Table 5.1 Coalition Order of Battle

---

[213] *Conduct of the Persian Gulf War: Final Report*, 114.
[214] "Forces Committed," *Military Review*: 80-81.
[215] Norman Friedman, *Desert Victory, The War for Kuwait* (Annapolis, MD: Naval Institute Press, 1991), 318.

Air Forces	591	Combat aircraft[216]
Ground Forces	540,000	Troops
	4,200	Tanks
	2,800	APCs
	3,100	Artillery pieces[217]
Naval Forces	10	Missile attack boats
	3	Subchasers
	7	Patrol boats
	8	Minesweepers[218]

Table 5.2 Estimated Iraqi Order of Battle

Planning the Offensive

Planning for an offensive to force Iraqi forces from Kuwait became a high priority beginning in early November. One of the initial steps was to identify Iraq's centers of gravity. They were perceived as being:
- Command, control, and leadership of Saddam Hussein's regime.
- Iraq's weapons of mass destruction, which included potential nuclear, chemical, and biological (NBC) warfare production facilities, as well as the means to deliver them.
- The Republican Guard.[219]
- After the centers of gravity were identified, planning became a step-by-step process, each succeeding phase building on the results of the previous one. In accordance with Secretary of Defense guidance, CINCCENT outlined his intentions in his "Mission Statement." CINCCENT's goal would be to conduct offensive operations to:
- Neutralize the Iraqi National Command Authority.
- Eject Iraqi armed forces from Kuwait.
- Destroy the Republican Guard.
- As early as possible, destroy Iraq's ballistic missile, NBC capability.
- Assist in the restoration of the legitimate government of Kuwait.[220]

---

[216] Ibid., 308.
[217] *Conduct of the Persian Gulf War: Final Report*, 113.
[218] Friedman, 361.
[219] *Conduct of the Persian Gulf War: Final Report*, 94.
[220] Ibid., 96.

The Gulf War Case

Based on the mission statement and President Bush's original national policy objectives, CINCCENT established theater military objectives as follows

- Attack Iraqi political-military leadership and C^2.
- Gain and maintain air superiority.
- Sever Iraqi supply lines.
- Destroy known nuclear, biological, and chemical production, storage, and delivery capabilities.
- Destroy Republican Guard forces in the Kuwaiti Theater of Operations (KTO).
- Liberate Kuwait City.[221]

The planning itself attempted to capitalize on Coalition military strengths and take advantage of Iraqi weaknesses. The CINCCENT "Concept of Operations" was developed taking into account this philosophy, as well as the stated military objectives. Based on a four-phase offensive campaign, the concept directed the component commanders to plan their operations in the following manner:

- Conduct a coordinated, multi-national, multi-axis air, naval, and ground attack.
- Focus a strategic air campaign on enemy centers of gravity.
  - Iraqi National Command Authority
  - NBC capability
  - Republican Guard forces command
- Progressively shift air operations to and conduct ground operations in the KTO to:
  - Isolate KTO and sever Iraqi supply lines.
  - Destroy the Republican Guard force.
  - Liberate Kuwait City with Arab forces.

The four phases of the war were identified as follows:

- Phase I -- Strategic Air Campaign
- Phase II -- Air Supremacy in KTO
- Phase III -- Battlefield Preparation
- Phase IV -- Offensive Ground Campaign.[222]

With the CINC's guidance in hand, the component commanders (Army, Navy, Air Force, Marine Corps, and Special Forces) developed

---

[221] Ibid., 96-97.
[222] Ibid., 98.

more detailed plans on how each would prosecute his part of the war. Their plans were essentially complete by mid-December, and the overall offensive campaign plan was subsequently approved by the Chairman of the JCS, the Secretary of Defense, and the President.

Desert Storm

The Air Campaign

Gulf lesson one is the value of air power. ... (it) was right on target from day one. The Gulf war taught us that we must retain combat superiority in the skies. ... Our air strikes were the most effective, yet humane, in the history of warfare.[223]

Planning for the air campaign started immediately after the invasion of Kuwait when CINCCENT requested the Air Staff in the Pentagon to develop a conceptual air offensive campaign against strategic targets in Iraq. Instead of the limited and incremental phasing of air power to "send signals" as was done in Vietnam, this plan called for the destruction of 84 targets in *Iraq* within one week.[224] The plan was refined throughout the Desert Shield buildup, and the planners were able to address the later stages of battlefield preparation in support of the ground offensive.

The Desert Storm air campaign plan was based on the President's initial objectives and derived its overall direction from CINCCENT's Concept of Operations. The Joint Force Air Component Commander (JFACC) staff identified the campaign's five major objectives, along with 12 target sets to accomplish them. The JFACC objectives and target sets included:

1. Isolate and incapacitate the Iraqi regime.

    Target sets included:
    - Leadership command facilities.
    - Crucial aspects of electricity production facilities that powered military and military-related industrial systems.
    - Telecommunications and C^3 systems.

2. Gain and maintain air supremacy to permit unhindered air operations.

---

[223] Ibid., 117.
[224] Ibid., 121.

Target sets included:
- Strategic Integrated Air Defense System (IADS), including radar sites, surface to air missiles (SAMs), and IADS control centers.
- Air forces and airfields.

3. Destroy NBC warfare capability.
Target sets included:
- Known NBC research, production, and storage sites.

4. Eliminate Iraq's offensive military capability by destroying major parts of key military production, infrastructure, and power projection capabilities.
Target sets included:
- Military production and storage sites.
- Scud missiles and launchers, production, and storage sites.
- Oil refining and distribution facilities, as opposed to long-term production capabilities.
- Naval forces and port facilities.

5. Render the Iraqi army and its mechanized equipment in Kuwait ineffective, causing them to collapse.

Target sets included:
- Railroads and bridges connecting military forces to means of support.
- Army units including the Republican Guard Forces Command (RGFC) in the KTO.[225]

Unlike Vietnam, few militarily significant targets were placed off limits, and those that were had to be justified on historical, religious, or cultural grounds, along with the desire to minimize collateral damage and casualties.

The air campaign was developed to support the four phases of the war, with the final effort focused on advancing the ground offensive. CINCCENT planners estimated Phases I-III would last approximately 18 days, during which Iraqi combat effectiveness in the KTO would be reduced by half.[226]

---

[225] Ibid., 125-126.
[226] Ibid., 135.

The Air War

When Operation Desert Storm was initiated on 17 January 1991 there were 2,430 Coalition fixed-wing aircraft (1,796 strike aircraft) in theater, enough to allow the Republican Guard divisions inside Kuwait—originally on the target list for Phase III—to be struck on the first day. The magnitude and details of the operations of that first night and succeeding days will be left to further study. The important point is that these early operations were exceedingly successful in gaining air superiority for the Coalition and in destroying strategic targets inside Iraq. In fact, only bad weather during the first week was effective in slowing down the air onslaught against Iraqi targets.

The actual "by week" air war objectives changed, as operations dictated, through the first five weeks of the offensive. During the first week the primary targets were strategic air defenses and C^3 networks, leadership infrastructure, NBC facilities, the Iraqi air force, electrical power grid, Scud missiles and sites, and Iraqi ground and naval forces in the KTO. In week number two emphasis continued on strategic targets and anti-Scud efforts, but the focus had already started to shift to lines of communication (LOCs) between Iraq and the KTO, as well as to Iraqi forces in the KTO. By the third week emphasis was definitely placed on the degradation of Iraqi forces in Kuwait through direct attack and destruction of the logistics supply system. By D+18 of the air war, Coalition intelligence believed that Iraq's ability to resupply forces in the KTO was below that required to sustain the forces during combat operations.[227]

Weeks four and five saw continued emphasis placed on attacking the Iraqi ground forces, both those along the Kuwait-Saudi border and the Republican Guard in the rear. Since the Coalition maintained air superiority and a significant segment of the Iraqi communications network had been destroyed, Iraq's intelligence capability was almost nonexistent by this time.

---

[227] Ibid., 179.

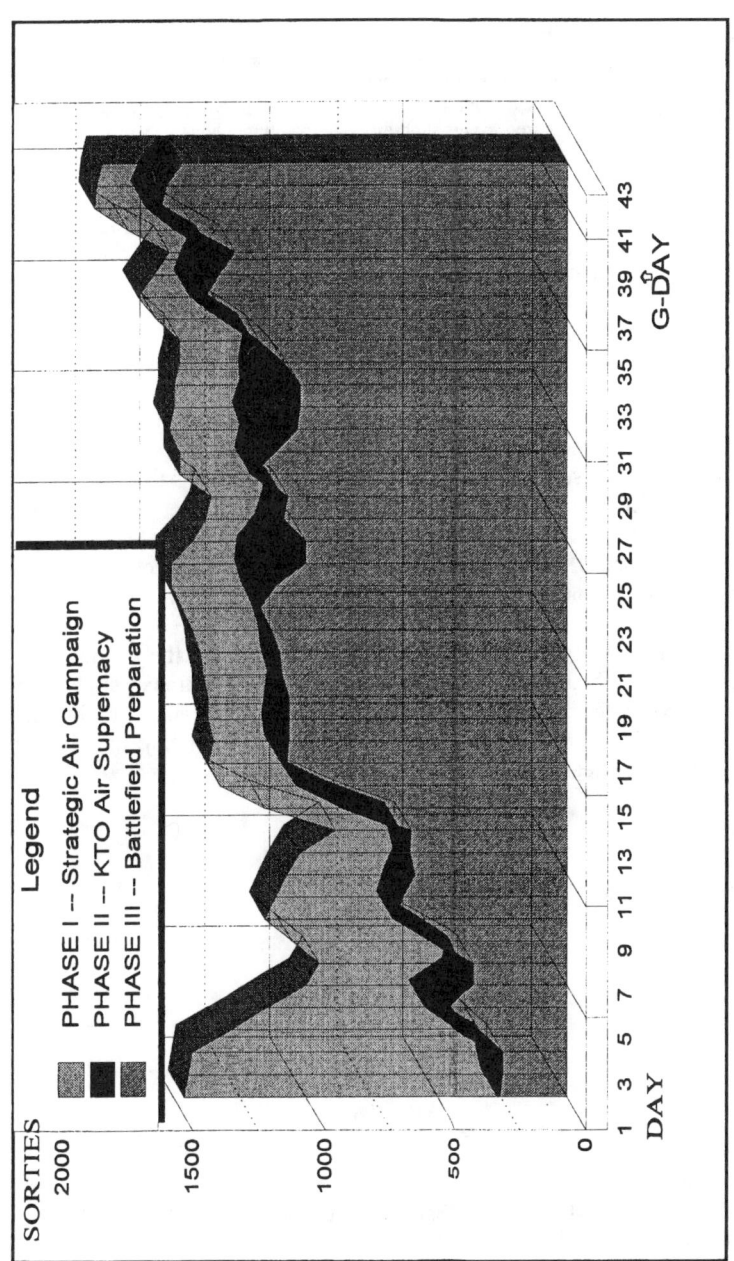

Figure 5.2. Sorties by Phase in the Air Campaign

Figure 5.2 [SOURCE: U.S. Department of Defense, *Conduct of the Persian Gulf War: Final Report to Congress* (Washington, DC: Department of Defense, April 1992), 135] tracks the number of air sorties targeted by phase. It is significant to note how quickly the emphasis shifted from strategic targets to battlefield preparation and how few resources were required for air supremacy support throughout the war. Figure 5.3 [SOURCE: U.S. Department of Defense, *Conduct of the Persian Gulf War: Final Report to Congress* (Washington, DC: Department of Defense, April 1992), 179] illustrates the remarkable effect this air interdiction had on Iraq's ability to resupply its forces in Kuwait. The chart is based on resupply movements from Baghdad to Al-Basrah near the Iraq-Kuwaiti border. By the time the ground war was initiated at D+38, Iraq's capability to logistically support the KTO was reduced to almost zero.

More than 35,000 attack sorties were flown against KTO targets during the war, including 5,600 against Republican Guard forces.[228] By G-Day, the start of the ground offensive, which corresponded to D+38, CENTCOM estimated that the effectiveness of Iraqi front-line forces had been reduced by 50 percent due to casualties, desertion, and lack of logistics support, all precipitated by the air campaign.[229] Figure 5.4 [SOURCE: U.S. Department of Defense, *Conduct of the Persian Gulf War: Final Report to Congress* (Washington, DC: Department of Defense, April 1992), 188] shows the estimated destruction of Iraqi ground equipment by Coalition air forces prior to G-Day.

The Iraqi air force was in even worse condition. Of the original 591 combat aircraft, 35 were shot down in air-to-air combat, 141 were destroyed in their shelters, and 148 were flown to Iran by their Iraqi pilots.[230] The rest were dispersed around the country, no longer presenting a threat to Coalition forces.

---

[228] Ibid., 179.
[229] Ibid., 191.
[230] Matthew M. Hurley, "Saddam Hussein and Iraqi Air Power," *Airpower Journal* 4 (Winter 1992): 12.

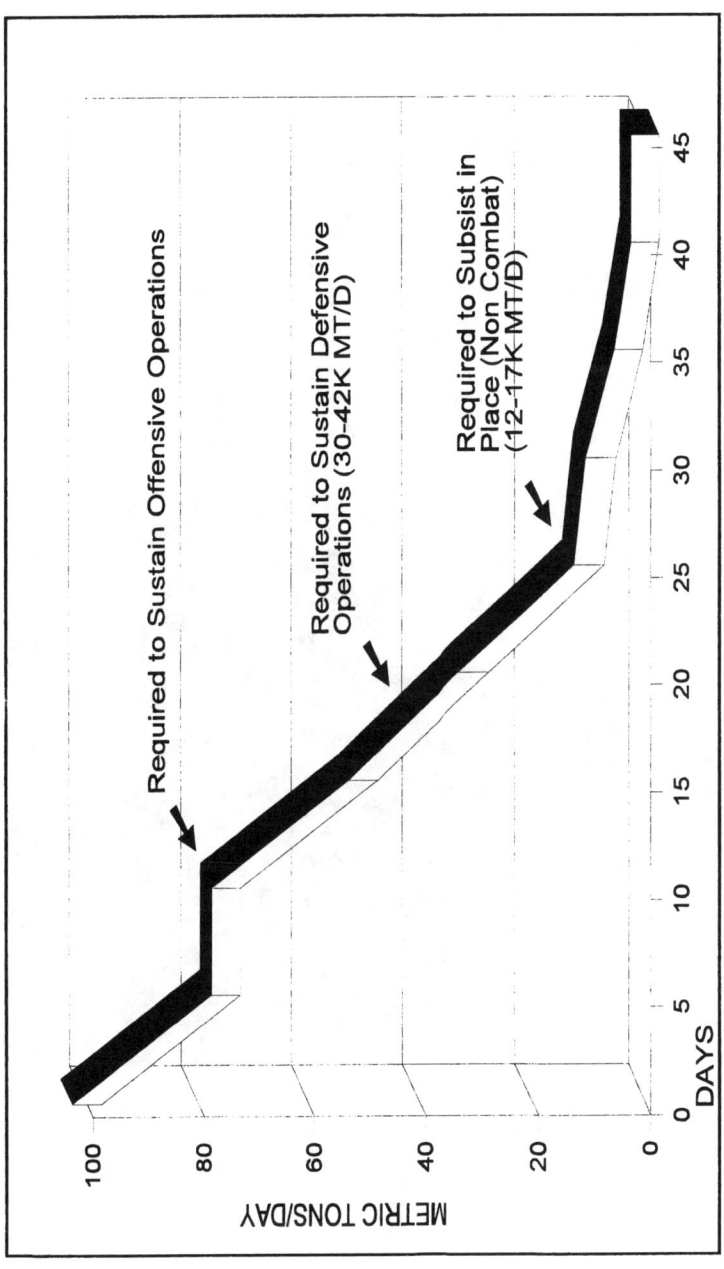

Figure 5.3 Iraqi Resupply Movements

118 *War as an Instrument of Policy*

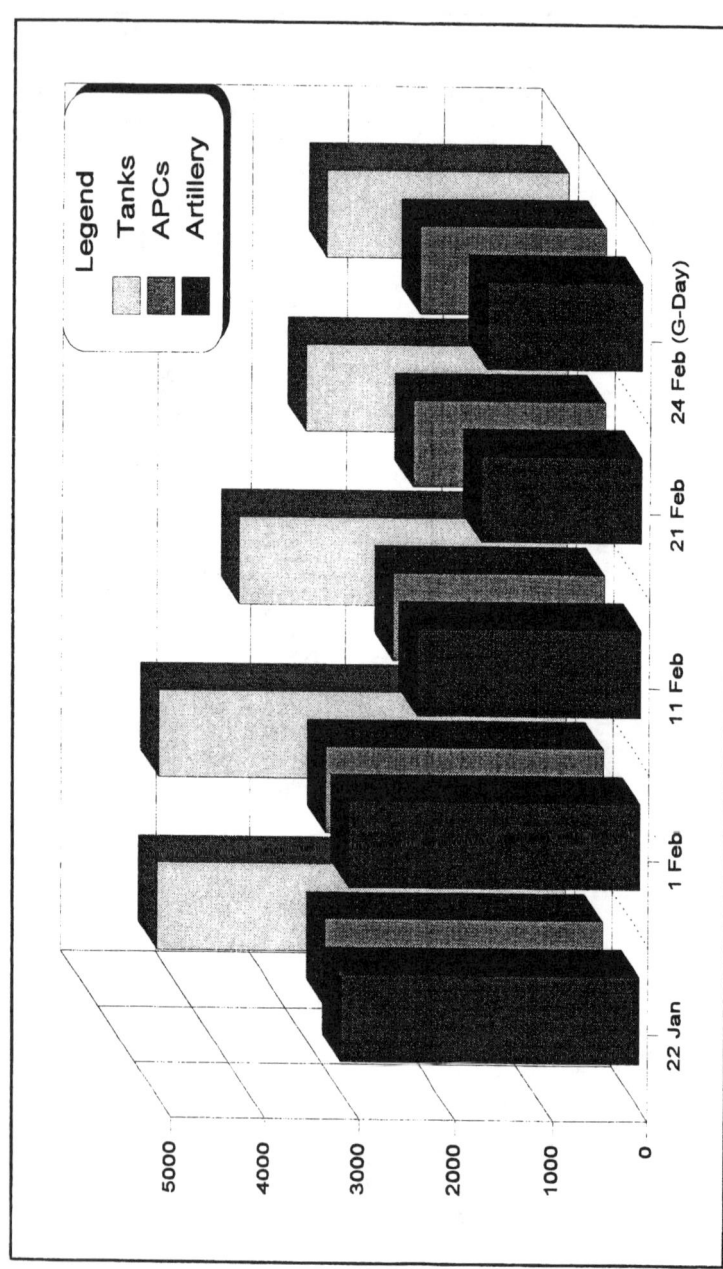

Figure 5.4. Iraqi Equipment Degradation in KTO Prior to G-Day

The Gulf War Case   119

The air campaign continued throughout Iraq and the KTO during the four-day Coalition ground offensive. Emphasis was placed on supporting ground forces through air interdiction and close air support. Primary targets included artillery, armor, armored personnel carriers, supply vehicles, command posts, and troops in defensive positions. The advance of Coalition ground forces was so rapid, however, that interdiction took a back seat to direct battlefield support.[231]

The Maritime Campaign

Throughout the Gulf conflict, the United States Navy deployed a total of 165 ships to the Persian Gulf and Arabian, Red, and eastern Mediterranean Seas. This armada included six carrier battle groups with their associated air wings. Coalition nations deployed another 65 ships in the region. As a result of this overwhelming combat force, control of the seas was never an issue.

To support CINCCENT's plan of an amphibious invasion, the following "major tasks" were developed for naval support of Desert Storm:

- Phases I and II -- Strategic Air Campaign and Establishment of Air Superiority

    - Conduct the air campaign.
    - Establish sea control and conduct mine countermeasures (MCM) operations in the northern Persian Gulf.
    - Attack shore facilities that threaten naval operations.

- Phase III -- Battlefield Preparation
    - Carry out Phases I and II tasks.
    - Attack Iraqi ground forces with aircraft and naval gunfire.

- Phase IV -- Offensive Ground Campaign
    - Continue to carry out Phases I, II, and III tasks.
    - Conduct amphibious feints and demonstrations in the KTO.
    - Be prepared to conduct an amphibious assault to link up with Marine Corps forces near Ash Shuaybah, between the Kuwait-Saudi border and Kuwait City.[232]

---

[231] *Conduct of the Persian Gulf War: Final Report*, 197.
[232] Ibid., 255-256.

Using an offensive antisurface warfare (ASUW) concept, the Coalition naval forces were able to detect and destroy Iraqi ships well before they could employ antiship weaponry. In fact, many of the 143 Iraqi ships damaged or destroyed were attacked in their ports, still alongside their piers.

Results of ASUW activity throughout Desert Shield/Storm can be summarized as follows:

- 143 [28 major] Iraqi naval vessels destroyed or damaged.
- All Iraqi naval bases/ports significantly damaged.
- All northern Persian Gulf oil platforms searched and secured.
- No attacks by Iraqi surface vessels against Coalition forces.[233]

Coalition naval forces secured the right flank of Coalition land forces during the ground offensive. This threat of an amphibious landing focused Iraqi attention on the beaches in the East rather than a Coalition advance in the western desert. Should an amphibious landing have been required, Iraqi mines, beach obstacles, and fortifications would probably have caused significant casualties in the landing force. Fortunately, this option was never exercised.

The Ground Campaign

With the ubiquitous media coverage of Desert Shield and Desert Storm, the launching of a ground offensive to retake Kuwait was probably one of the most anticipated events in the annals of modern warfare. Nevertheless, the Coalition deception plan worked almost to perfection, aided by the destruction of Iraqi intelligence capabilities. This success was one of the main reasons why the ground war concluded with an overwhelming military victory.

---

[233] Ibid., 268.

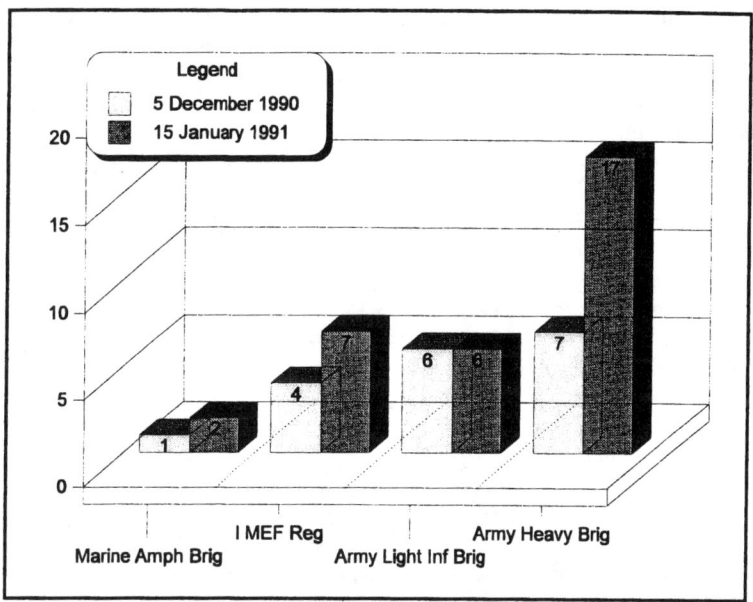

Figure 5.5 Arrival of U.S. Ground Forces

SOURCE: Department of Defense, *Conduct of the Persian Gulf War: Final Report to Congress* (Washington, DC: Department of Defense, April 1992), 321.

CINCCENT briefed his Concept of Operations for the ground war to his commanders on 14 November 1990. His strategy and intent were to "maximize friendly strength against Iraqi weakness and terminate offensive operations with the RGFC destroyed and major US forces controlling critical Lines of Communication (LOCs) in the Kuwaiti Theater of Operations."[234] The Operation Desert Storm OPLAN directed that the ground campaign be conducted in four phases:

- Phase I -- Logistical buildup
- Phase II -- Force repositioning
- Phase III -- Ground attack
- Phase IV -- Tactical consolidation.[235]

Significant numbers of U.S. ground forces, particularly those units with heavy armor, did not arrive in Saudi Arabia until the last month of Desert Shield. (See Figure 5.5.) This limited crucial desert training

---

[234] Ibid., 317.
[235] Ibid., 318.

before the initiation of the ground war. Phase I of the plan did not, in fact, coalesce until the day prior to launching the offensive, when elements of VII Corps, the principal unit tasked with the critical offensive drive in the West, moved from Saudi ports to the forward attack position.

Logistical support for the ground offensive was substantial. ARCENT (U.S. Army), along with British and French forces, numbered 258,701 soldiers, 11,000+ tracked vehicles, 47,000 wheeled vehicles, and 1,619 aircraft.[236] To support this force on G-Day, 29.6 million meals, 36 million gallons of fuel, and 114.9 thousand tons of ammunition had to be positioned in the West, while not alerting Iraqi forces to the shift of Coalition forces in that direction.[237] Supply bases contained enough equipment and stores to support combat operations for up to 60 days.

ARCENT operations plans were grounded on AirLand Battle doctrine, which "is centered on the combined arms team, fully integrating the capabilities of all land, sea, and air combat systems, and envisions rapidly shifting and concentrating decisive combat power, both fire and maneuver, at the proper time and place on the battlefield."[238] The success of this concept was predicated on the ability to effect the following aims:

- Initiative -- to set or change the terms of battle by offensive action.
- Agility -- The ability of friendly forces to act mentally and physically faster than the enemy.
- Depth -- The extension of operations in space, time, and resources.
- Synchronization -- The arrangement of battlefield activities in time, space, and purpose to produce maximum relative combat power at the decisive point.[239]

The final ground offensive plan involved a Coalition attack across the entire front along the Kuwait-Saudi Arabian border. The primary effort would take place in the West, well coordinated with offensive drives nearer the sea.

> The main attack was designed to avoid most fixed defenses, drive deep into Iraq, envelop Iraqi forces from the west and attack and destroy Saddam Hussein's strategic reserve—Republican Guard

---

[236] Ibid., 333.
[237] Ibid., 333-334.
[238] Ibid., 329.
[239] Ibid., 329-330.

armored and mechanized infantry divisions augmented by several other Iraqi Army heavy divisions. This wide left sweep, sometimes referred to as the "Hail Mary" plan, emphasized the key tenets of Airland Battle doctrine.[240]

Once the air campaign achieved its objective of cutting the Iraqi combat capability by 50 percent, the ground war was launched. It began on 24 February 1991 and ended 100 hours later. During that brief but highly intense time, Coalition forces used deception, surprise and maneuver to avoid Iraqi fixed defenses and to apply Coalition strength against Iraqi weakness.

Members of the Coalition ground force conducted attacks all along the Kuwaiti-Saudi border to fix Iraqi forces and make it appear the primary Coalition offensive would be conducted near the coast and directed at the heart of the Iraqi defenses. Meanwhile, the main Coalition attack, executed primarily by the U.S. Army's VII Corps, circumvented Hussein's fixed defensive line along the border and made a wide, sweeping drive around the Iraqi right flank in the West engaging the Republican Guard, Iraq's strategic reserve. "This sweep employed the strength of AirLand battle doctrine, including agility, depth, synchronization of combat power, initiative, and sustainment of the force."[241] The threat of a Marine amphibious assault along the Kuwaiti coast prevented reinforcement of the Republican Guard.

Iraqi defensive lines in the East were ultimately breached, after which Coalition forces pushed on to Kuwait City. At the same time, Republican Guard units encountered on the battlefield were nearly destroyed and virtually all in the KTO were cut off from retreat.

The ground offensive ended at 0800 on 28 February. All CINCCENT objectives had been achieved, and Coalition forces:

- Controlled the critical lines of communications in the KTO.
- Ejected the Iraqi forces from Kuwait.
- Secured the Kuwait International Airport and crossroads west of Kuwait City.
- Flanked, cut off, and destroyed Republican Guard forces.
- Liberated Kuwait City.[242]

The final victory was one of the most one-sided in the annals of warfare. The number of Iraqi losses underline this fact. CENTCOM's estimate of the Iraqi losses is presented in Figure 5.6.

---

[240] Ibid., 338.
[241] *Conduct of Gulf Conflict: Interim Report*, 2-7.
[242] *Conduct of Gulf War: Final Report*, 411.

War Termination

Even though all of the original political objectives declared by President Bush had been achieved, events that occurred in the aftermath of the Gulf War were disappointing with respect to war termination. Kurds were slaughtered by the thousands at the hands of remnants of the Iraqi army. Iraq only grudgingly acquiesced to United Nations demands to inspect NBC and Scud facilities. And, most disturbing, Saddam Hussein remained in office.

Although not explicitly identified as one of the original political objectives, the removal of Saddam Hussein from office must have been high on the *wish list* of fallout resulting from an Iraqi defeat. However, the war was stopped short of sending Coalition forces into Baghdad, and the hoped-for Kurdish rebellion never materialized into a formidable insurrection. In addition, several revolts and assassination plots against Hussein involving the Iraqi military were unsuccessful. In the end, Hussein appeared to be more strongly in control as the "president" of Iraq than ever.

Almost immediately following the end of the war, Hussein started to rebuild his army. Of the 60 to 70 divisions Iraq had prior to the war, 25 to 30 never saw action in Kuwait, including 5 Republican Guard divisions.[243] An estimated 700 tanks survived the war and became the nucleus of new Iraqi armored divisions.[244] With this combined force, Hussein was able to assemble an army that remains one of the strongest in the Middle East and contributes to his status as one of the most important political leaders in the region.

---

[243] *Triumph Without Victory*, 412.
[244] Ibid., 412.

The Gulf War Case

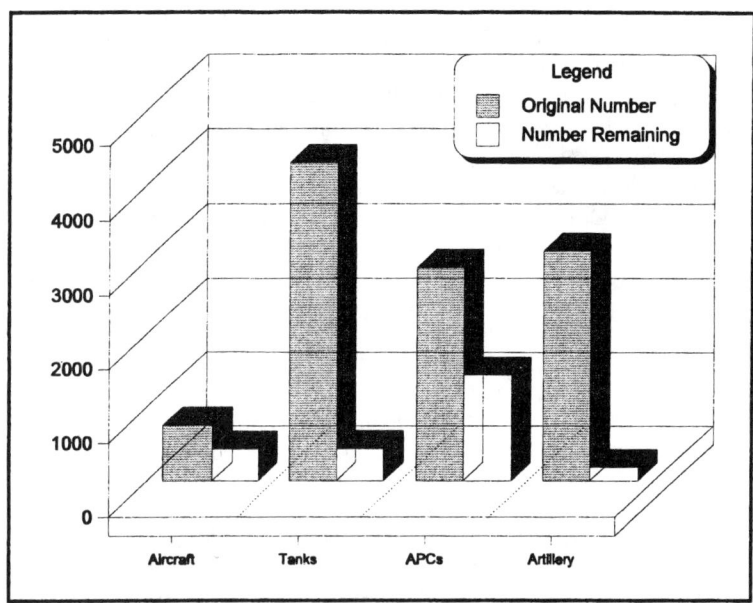

Figure 5.6 Iraqi Losses During the Gulf War

SOURCE: Department of Defense, *Conduct of the Persian Gulf War: Final Report to Congress* (Washington, DC: Department of Defense, April 1992), 204, 411.

In the application of the principles of war, could decisions have been made to bring about a more satisfactory conclusion to the conflict? An examination of events leading up to the cease-fire and its aftermath indicates they could have by revealing two critical errors. The first was the decision to end the war prior to the total destruction of the Republican Guard. The second was the decision to allow Saddam Hussein to use air power to counter uprisings by the Iraqi people.

In the application of the principles of war, doctrine calls for "exploitation" when the advantage is attained, the phase of combat that makes possible the expansion and consolidation of gains, keeps the enemy off balance, and leads to the acquisition of tactical and strategic objectives. In the Gulf War the Coalition forces were unquestionably in the exploitation phase when a cease-fire was declared at the 100-hour mark of the ground campaign. Significant numbers of Iraqi troops were allowed to escape certain destruction as a result. These same Iraqi forces were used a few days later by Hussein to crush rebel uprisings whose primary purpose was to overthrow the Iraqi dictator, a rebellion the

United States wanted to succeed. Ironically, the United States also failed to exploit a significant military advantage over the Viet Cong after the 1968 Tet offensive in the Vietnam War.

Haunted by this legacy, the allies at first seemed resolved to hold Hussein to the cease-fire agreement that prohibited him from using fixed-wing aircraft as a weapon against the Kurds. However, after shooting down two of his planes that were in flagrant violation of the cease-fire conditions, Coalition forces later allowed others, along with helicopters, to fly against the rebels with impunity. According to President Bush, the United States was not going to become involved in an internal conflict à la Vietnam.

This decision, along with the incomplete destruction of the Republican Guard, proved critical in allowing Hussein to defeat the rebel uprising. But even this course of events must be examined in the strategic context of conflict termination and stability in the region. For instance, what would have happened if the Iraqi rebellion had succeeded and Hussein was forced from power? Could the two factions confronting the government have worked together to form a strong central government? History reveals that this would have been difficult at best.

If Hussein had been deposed, would Iran or Syria have attempted to step into the vacuum to enhance their own claim to power in the region? If nothing else, a strong Iraq in the past had been a stabilizing influence on the aspirations of other Middle East nations. Perhaps Saddam Hussein, without the backing of a large military force, would, in the long run, be the lesser of two evils.

Chapter 6

The South African Case

Angola - The Country

Angola's geographic location and features must be fully understood to gain a thorough appreciation for strategic and operational decisions made during the Angolan War. For example, the fact that Angola is located a significant distance from the South African border played a major role in how Pretoria elected to counter the perceived threat and on how long logistics lines impacted Pretoria's ability to support its force in the field. Figure 6.1 shows the geographic location of African countries directly involved in the conflict.

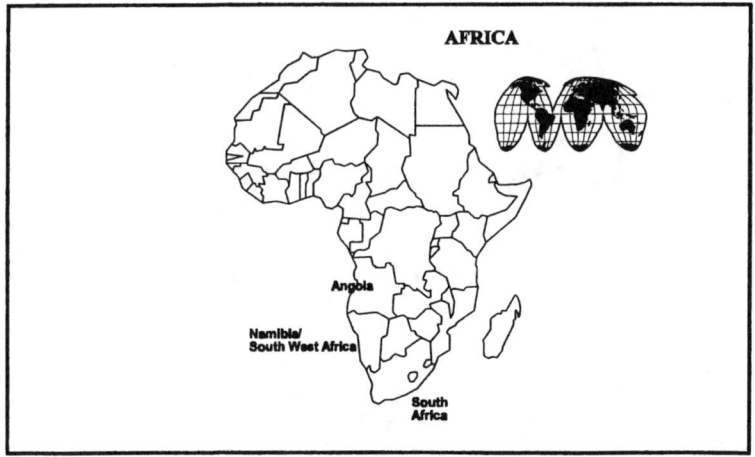

Figure 6.1 The African Continent

Angola is a fairly large country in area, but has a small population in relation to its size. The country contains 1,246,700 square kilometers (481,354 square miles), making it slightly less than twice the size of Texas. The population in 1991 was estimated to be approximately 8.7 million, an average of only 7 persons per square kilometer (2.6 per square mile).[245]

Angola's low population density was an important factor in the war because it gave invading forces the capability to move throughout the rural areas with relative impunity. In one incursion, the South African Defence Force was able to penetrate as far as 500 kilometers into Angola without encountering any local inhabitants.[246] Relevant cities, towns, and geographic features key to the conduct of the war are depicted on the map in Figure 6.2.

Figure 6.2 Map of Angola

---

[245] Central Intelligence Agency, *The World Fact Book 1991* (Washington, DC: U.S. Government Printing Office, 1991), 8.

[246] Brigadier W.P. Sass, interview by author, 27 July 1993, Pretoria, South Africa.

Most of Angola's inland area consists of an expansive high plateau that descends abruptly to a narrow plain along the coast. The highland area supports a vacillating climate and terrain features that severely impacted the conduct of the war. The climate, which includes rainy/dry seasons, had to be considered when planning the military campaigns. The intense heat sapped soldiers' strength, while the rainy season provided a breeding ground for insects that spread many debilitating diseases. Terrain features, particularly rivers, contributed directly to the outcome of battle between opposing forces. Forty-three percent of the country is also covered by forests and woodlands. This "bush" provided opposing armies with excellent concealment for guerrilla-type operations and significantly impeded the mobility of conventionally designed forces.

Angola also boasts significant reserves of petroleum and large deposits of diamonds, iron ore, phosphates, copper, bauxite, and uranium. This mineral wealth, as well as the country's agricultural potential, heightened the interest of outside powers, particularly the communist bloc, in what was at first an internal civil war.

Prelude to Invasion

Angola Moves Toward Independence

South Africa's strategic interest in Angola dates back to the early 1960s when warfare broke out between Portugal and indigenous Angolan guerrillas intent on winning independence for their country. This conflict, along with revolutionary movements in other Portuguese colonies, ultimately led to a coup by the Portuguese military in April 1974 and the introduction of an independent government in Angola the following year. Ensuing revolutionary activity against the newly formed Angolan government, aided by external political, economic, and military involvement, compelled the South African government to take action in support of what it considered to be a threat to its northern border and a heightened risk to its own internal stability.

Dr. Willem Van der Waals, a retired brigadier from the South African Defence Force with intimate, first-hand knowledge of the Angolan insurgency, traced intensified South African concerns in Angolan affairs to as early as 1960. According to Van der Waals, the SADF concluded and documented in November of that year that an independent Angola would become a threat to South Africa: "If the white-controlled states to the north of the RSA [Republic of South

Africa] should become independent, the potential enemy would be on the RSA's borders."[247]

By 1966 rebel activity from Angola was already spilling over into South African-controlled South West Africa (SWA). The SADF's Chief of Defence Staff determined that "the RSA is the ultimate and major target of the OAU[248] and their communist supporters. The likelihood of revolutionary war in the RSA will depend upon the successes achieved by the rebels in adjoining territories."[249] As a result, limited South African military cooperation with Portugal widened in 1970 and continued to expand through 1974 when the security situation in Angola was at its worst. Although much of the detailed information is still classified, it is known that SADF and South African Police (SAP) operations in Angola during this time included the hot pursuit of rebels across the Angolan border, limited air support, aerial survey and mapping, assistance with road construction, and medical and food supply.[250]

The 1974 coup in Portugal assured autonomy for Angola, but by the time independence was achieved on 11 November 1975, the country was embroiled in a civil war between three rival liberation groups, each supported by foreign military arms and advisors. The *Movimento Popular de Libertaçao de Angola* (MPLA), a largely urban-based revolutionary group, was situated in and around Luanda, the Angolan capital, and derived most of its support from the Soviet Union and Cuba. The *Frente Nacional de Libertaçao de Angola* (FNLA) garnered its support from the people of northern Angola, and relied on aid from the United States and China and military assistance from Zairean army units. The *União Nacional para a Indepéndencia Total de Angola* (UNITA) established its base for resistance in southern Angola where it was initially supported by the United States and South Africa.

In early 1975 Henry Kissinger, then Secretary of State, approved a Central Intelligence Agency (CIA) request to provide covert assistance to the FNLA in an effort to counter Soviet expansionism in southern

---

[247] Willem S. Van der Waals, *Portugal's War in Angola, 1961-1974* (Rivonia, South Africa: Ashanti Publishing (Pty.) Ltd., 1993), 133, citing SADF Archives.

[248] The Organization of African Unity is an organization established in 1963 by the Conference of Heads of African States whose primary objective was the liberation of African states from all forms of colonial rule.

[249] Ibid., 133.

[250] Ibid., 208-209.

Africa.[251] This was part of a Western program to recruit European mercenaries to fight against the MPLA and encourage the South African Defence Force to enter the Angolan War in support of the FNLA and UNITA. The Soviet Union retaliated in turn by providing arms shipments to the MPLA. This group was also aided by several hundred Cuban "advisors."

In July 1975, U.S. President Gerald Ford approved a total of $31.7 million to supply arms to the FNLA and UNITA.[252] This was the maximum allowed under U.S. law without approval from Congress. Despite the infusion of U.S. military aid however, MPLA forces gained control of the capital city and rapidly extended their influence over large sections of the remainder of the country.

South Africa watched the events in Angola with increasing concern. As it did, the Marxist-dominated MPLA made significant gains in the civil war, aided by the arrival in Luanda of substantial amounts of Soviet weapons and large numbers of Cuban troops. Finally, South African Minister of Defence P.W. Botha could watch no longer. In September 1975 the SADF kicked off Operation Savannah with a three-pronged attack across the SWA/Angolan border.

The operation was initiated to prevent a Marxist-oriented organization (sympathetic to guerrilla groups opposed to South Africa) from taking control over the government of a country that had significant strategic importance.[253] Pretoria's original intent was to secure as much traditional UNITA territory as possible prior to Angolan independence on 11 November and then to withdraw.[254]

The operation itself initially exceeded everyone's expectations as the South African force rapidly moved 500 miles into Angola. Luanda itself was within striking distance, but to take the city would have required a larger investment of manpower and equipment with an associated increase in casualties; a commitment the South African government was unwilling to make. In mid-December the Cabinet

---

[251] Sanford J. Ungar, *Africa; The People and Politics of an Emerging Continent*, 3rd rev. ed. (New York: Simon & Schuster, Inc., 1989), 72.

[252] Fred Bridgland, "Angola and the West," in *Challenge: Southern Africa Within the African Revolutionary Context*, ed. Al J. Venter (Gibraltar: Ashanti Publishing Limited, 1989), 129.

[253] Jeremy Grest, "The South African Defence Force in Angola," in *War and Society: The Militarisation of South Africa*, ed. Jacklyn Cock and Laurie Nathan (Claremont, S.A.: David Philip, Publisher (Pty.) Ltd., 1989), 122.

[254] Fred Bridgland, *Jonas Savimbi: A Key to Africa* (Johannesburg: Macmillan South Africa (Publishers) (Pty.) Ltd., 1986), 145.

made the decision to withdraw from Angola while leaving as much territory as possible in the hands of Jonas Savimbi's UNITA forces.

South Africa in Isolation

The SADF's performance on the battlefield notwithstanding, the overall South African Angolan operation became self-defeating. First, the intervention escalated the conflict and provided justification for the MPLA to request additional Cuban troops. In six months, the number of Cubans in-country supporting the MPLA more than tripled, from 4,000 in December 1975 to approximately 13,500 by the spring of 1976.[255] Second, the intervention encouraged uncommitted African states, including Nigeria, the most powerful country in central Africa, to formally recognize the MPLA as the legitimate government of Angola. Third, the invasion accented American covert involvement in the region, which led to an about-face in U.S. policy. Finally, when the operation was disclosed to the world, the international community condemned South Africa for its aggression and doomed UNITA as a recipient of South African aid.

Operation Savannah had its internal repercussions as well. Initiated without the knowledge of Parliament, it highlighted a *modus operandi* of the Pretoria government that had dominated the nation for the past decade. In addition, censorship of the press prevented the South African public from knowing of the SADF's involvement in the war until after the operation had been completed. Finally, because then Prime Minister Vorster was concerned about international reaction to South African involvement, the SADF commitment was purposely limited in scope, to the detriment of providing proper support to the troops in the field.

Operation Savannah was significant for two other important reasons. First, it underscored the lack of commitment on the part of the United States to support groups and ventures in sub-Sahara Africa. And second, it emphasized the untenable position the South African government was in with respect to both internal and external concerns.

In the case of the United States, the South African government claimed, and substantial evidence is available to support the assertion, that the United States urged South Africa to enter the war in support of the FNLA and UNITA. In fact, Bernard Nossiter reported in the *Washington Post* that details on American support for South African involvement in Angola were delivered to South African Ambassador to

---

[255] Harold D. Nelson, ed., *South Africa: A Country Study*, Area Handbook Series (Washington, DC: U.S. Government Printing Office, 1981), 306.

the UN Pik Botha and by then U.S. UN Ambassador Daniel Patrick Moynihan. They were also supplied to Prime Minister Vorster by U.S. Ambassador to South Africa William Bowdler.[256] When knowledge of the South African Angolan operation became public, the U.S. Congress cut off all covert CIA aid to UNITA and the FNLA. Believing that it could not fight alone against ever increasing numbers of Cuban troops and expanding Soviet logistics support, Pretoria's leaders became extremely bitter that the West failed to give South Africa open support when the "chips were down." P.W. Botha pronounced:

> If America had shown her teeth the Russians might have capitulated, because they don't like to fight away from their fatherland. It is true the Cubans were there, but when they came up against us they ran away. The Americans' poor showing resulted in the majority of the OAU joining the strongman. In the future Angola will be remembered as the Free World's great lost opportunity.[257]

Seventeen years after the ignominious conclusion of Operation Savannah, officers in the SADF still point out how the United States deserted South Africa and left it to stand alone in supporting the struggle against a communist takeover in Angola.[258] Some of the more intellectually inclined officers, although convinced of the need to support the guerrilla forces, also criticize the South African government for accepting U.S. assurances of support.[259]

The second reason Operation Savannah was important to South Africa was because it demonstrated to the government and the SADF just how vulnerable the country was to both external and internal events. In short, the republic's so-called allies had forsaken it in the heat of battle for political reasons, leaving South Africa as more of a pariah state than ever. At the same time, the SADF discovered that it had a great deal of equipment and logistics limitations, while a large, modern army supported by the Soviet Union now dominated South

---

[256] *Washington Post*, 3 February 1976.

[257] Willem Steenkamp, *South Africa's Border War: 1966-1989* (Gibraltar: Ashanti Publishing Limited, 1989), 60.

[258] The author has had extensive conversations with members of the SADF and other specialists at numerous social and professional functions during the course of this research concerning the role the United States played in urging South Africa to become involved in Angola. Those who were personally involved at the time tend to display a more pronounced resentment than do younger South Africans.

[259] Brigadier Willem S. Van der Waals (SADF, Ret.), interview by author, 9 July 1994, Pretoria, South Africa.

West Africa's northern border. In addition, neighboring countries now controlled by Marxist regimes soon began training South West Africa People's Organization (SWAPO) guerrillas and exiled members of the African National Congress (ANC) to infiltrate SWA and South Africa itself to undertake revolutionary warfare. Meanwhile, Fidel Castro and Leonid Brezhnev began vocalizing what South Africa presumed would be a communist goal; the spread of a Marxist revolution throughout southern Africa, as far south as Cape Town.[260]

South Africa and Angola: 1976-1984

The Alliances

South Africa's Adversaries

The opposition South Africa faced during the Angolan War came from countries and organizations dominated by Marxist ideology. Although allegiances shifted at times, by 1975 those factions opposing South Africa included the Soviet Union, Cuba, Angola (MPLA), the African National Congress, the South West African People's Organization, and to a lesser extent, several other Eastern-bloc and Third World countries. These opponents all had their own reasons for involvement, roles in the conflict and relevance to the outcomes.

The Soviet Union. While the United States was constrained to providing approximately $32 million worth of arms to anti-Marxist forces in Angola prior to Congress cutting off covert funding, the Soviet Union poured in over $400 million worth of military equipment to its MPLA and Cuban allies.[261] As fighting escalated between the MPLA and UNITA over the next 10 years, Soviet aid and involvement increased. (See Table 6.1.)

---

[260] Fred Bridgland, *The War For Africa: Twelve Months that Transformed a Continent* (Gibraltar: Ashanti Publishing Limited, 1990), 12.

[261] John Stockwell, *In Search of Enemies: A CIA Story* (New York: W.W. Norton & Company, 1978), 180.

Year	Level of Aid
1956 to 1974	$63 million[a]
1977 to 1982	1 billion
1982 to 1984	2 billion
1986 to 1987	1 billion
1987 to 1988	1 billion

Table 6.1 Soviet Military Aid to Angola 1956-1987

SOURCE: Michael H. Armacost, "Regional Issues and U.S.-Soviet Relations," U.S. Department of State, Current Policy No. 1089, June 22, 1988, 5; quoted in Peter Vanneman, *Soviet Strategy in Southern Africa* (Stanford: Hoover Institution Press, 1990), 47.

[a]Figure includes nonmilitary aid.

When Congress cut off funding and the United States abdicated its position as a major player in Angolan affairs in early 1976, the door was left open for unimpeded Russian support for the MPLA. Soviet IL-62 jet transport planes ferried men and equipment from the Soviet Union and Cuba. MPLA troops were taken to the Soviet Union for training. The number of Soviet advisors in-country increased significantly. The MPLA was provided with increasingly more sophisticated weapons, including aircraft. The net result of this substantial support was that the FNLA quickly disintegrated into an ineffective organization and the Savimbi-led UNITA resistance group was driven into the bush.

When Savimbi reemerged from the bush in the late 1970s and started taking control of more and more Angolan territory, the Russians stepped up their involvement. In 1979 they attempted to demonstrate their resolve by sending a powerful naval task force, including an aircraft carrier, around the Cape of Good Hope. The task force stopped in Luanda to signal a strong commitment to the MPLA regime. In the early 1980s, SAM-6 anti-aircraft missiles and radars, SAM-9s, and Mig-23s were provided to the MPLA, and Cuban, Soviet, and perhaps East German pilots started flying Angolan combat aircraft in support of the ground forces. Finally, Soviet advisors started accompanying MPLA forces in their offensive operations against UNITA.

The Societ Union's massive military support for the MPLA along with Cuban forces entering the country as a surrogate army significantly escalated the contest between East and West for influence in sub-Sahara Africa.

> From the point of view of Washington, the Soviet-Cuban intervention in Angola was a dangerous and unprecedented power play, an offensive thrust that went beyond the traditional geographical perimeter of Russian interests and beyond the accepted postwar rules of the game for international behavior.[262]

The amount of support provided by the Soviet Union in men, material, logistics, and money significantly escalated the contest between East and West in sub-Sahara Africa. It was also indicative of the Soviets' ambition to strengthen their influence in the area. Arthur Klinghoffer, a professor of political science at Rutgers University, suggested that Soviet objectives in Southern Africa during the late 1970s and early 1980s were influence-driven and included the following:

- Extend Soviet logistic facilities, including the prepositioning of equipment and possibly even military personnel.

- Reduce Western political influence in the region.

- Obtain stronger African backing for Soviet international policies.

- Establish close ties with the most politically compatible states.

- Weaken Chinese influence.

- Support liberation movements opposed to colonialism or white-minority rule.

- Develop an image as a strong proponent of Third World efforts directed against colonialism and neocolonialism.

- Establish African political systems based on Marxist-Leninist ideology and Leninist party structure.

---

[262] Bruce D. Porter, *The USSR in Third World Conflicts*. (Cambridge, U.K.: Cambridge University Press, 1984).

- Extend political influence to give the Soviet Union some control over domestic African policy.[263]

Cuba. General Rafael del Pino Diaz defected from Cuba to the United States in May 1987 while he was deputy commander of the Cuban Air Force. During his debriefings at a Virginia site near Washington, D.C., he provided unique insight into the Cuban strategic concept for operations in Angola.

According to del Pino, at the time of Angolan independence in 1975, Fidel Castro correctly assessed the political and military environment of the southern African region. Apparently, as Cuba became more deeply committed to the Angolan civil war, Castro believed that the United States would not intervene directly because the Americans were just emerging from their Vietnam debacle. Western European countries would not become involved in any serious way, he also surmised, because they were too dependent on the United States for military action. Thus, the U.S.-supported FNLA would easily be defeated on the battlefields of northern Angola. And while the South Africans posed a serious military threat, Castro believed that once knowledge of their activity in Angola became known to the Western countries, intense pressure would force them to withdraw in the face of vehement condemnation.[264]

Based on these optimistic expectations, Castro relegated Cuba to the role of military proxy for the Soviet Union in Angola. For the next 13 years Cuban troops formed the underpinnings of military support for the MPLA and shouldered an ever increasing share of the fighting, with an associated increase in casualties as the conflict intensified. Although Cuban troops suffered losses in the early stages of the war, none of the Cubans were assigned to combat units between 1977 and 1983.[265]

Requirements, however, were about to change. As Jonas Savimbi's UNITA guerrilla units became more formidable in the early 1980s (with South African aid), the Soviets were obliged to launch a more proactive military strategy that bordered on conventional warfare. They

---

[263] Arthur Jay Klinghoffer, "Soviet Union and Superpower Rivalry," in *African Security Issues: Sovereignty, Stability, and Solidarity*, ed. Bruce E. Arlinghaus (Boulder, CO.: Westview Press, 1984), 27.

[264] General Rafael del Pino Diaz, interview by Fred Bridgland, December 1987, Washington, DC, quoted in Fred Bridgland, "Angola and the West," in *Challenge: Southern Africa Within the African Revolutionary Context*, ed. Al J. Venter (Gibraltar: Ashanti Publishing Limited, 1989), 132.

[265] Peter Vanneman, *Soviet Strategy in Southern Africa* (Stanford: Hoover Press, 1990), 51.

provided a substantial increase in military equipment and elected to station advisors with fighting units rather than requiring them to remain on a headquarters staff. At the same time, the Cubans made a commitment to assign soldiers directly to combat units. This was a critical decision because it ultimately required more than 400,000 Cuban soldiers to rotate through Angola, and resulted in more than 10,000 of these getting killed, wounded, or lost in action.[266]

Popular Movement for the Liberation of Angola. Opposition to Portugal's colonial rule in Angola began in the late 19th century. Organized resistance, however, did not evolve until the 1950s and 1960s when three nationalistic opposition groups were formed, the UPA, the MPLA, and UNITA.

The first, and least effective of the resistance organizations, was the *União das Populações de Angola* (UPA). Formed in the late 1950s under the leadership of Holden Roberto, the UPA tended to be pro-Western in its ideology because of Roberto's background in the international arena. The group's primary objective was the independence of Angola from Portugal.

Shortly after a limited armed struggle broke out against Portuguese rule in 1961, the UPA was assimilated into the Front for the National Liberation of Angola (FNLA), also led by Holden Roberto. The FNLA ceased being a viable entity in the political struggle for government control shortly after Angolan independence when American military aid was cut off. The military wing of the FNLA was subsequently defeated by Cuban troops allied with the MPLA. Many of the FNLA soldiers, having been members of the MPLA's Eastern Revolt led by Daniel Chipenda, later turned up as refugees on the Namibian border where they were integrated into the South African Defence Force as a new battalion, 32 Battalion.[267]

The FNLA's primary opponent was the MPLA. Founded in 1956, the MPLA was dedicated to defeating Portuguese colonialism and imperialism through an all-out struggle.[268] Prompt action by the Portuguese government forced the leftist-leaning group to go underground, but by 1965 the MPLA began to reemerge. The OAU provided an increasing amount of assistance. The Soviet Union also

---

[266] Fred Bridgland, "Angola and the West," in *Challenge: Southern Africa Within the African Revolutionary Context*, ed. Al J. Venter (Gibraltar: Ashanti Publishing Limited, 1989), 141.

[267] Brigadier W.P. Sass, interview by author, 27 July 1993, Pretoria, South Africa.

[268] Van der Waals, 48.

furnished active and exclusive support while helping to discredit Roberto of the UPA as an "American puppet."[269] Shortly after a visit by Che Guevara, Cuba established itself as a new source of assistance to the MPLA subversion effort. By 1966, MPLA power and influence had grown to the point where it found itself the dominant revolutionary organization in Angola.[270] Although the organization's fortunes cycled during the armed resistance years of 1962 through 1974, by 1975 the *Forças Armadas de Libertaçao de Angola* (FAPLA, the armed wing of the MPLA), with indispensable help from the Cuban army, claimed victory over the other two resistance movements, allowing the MPLA to take control of the newly independent government.

Following Angolan independence, the MPLA concentrated on consolidating its power. Critics from inside the organization were purged from the government. The loose anti-colonial coalition was transformed into a Marxist-Leninist party with all the trappings of closely guarded political power, socialist ideology, and state-structured institutions.[271] The nucleus of the MPLA economic policy was founded on central state ownership of productive resources and total control of the economy,[272] while the foundation of Angolan foreign policy was based on a close and supportive relationship with the Soviet Union, Cuba, and other Soviet bloc countries. Although Angola made some inconsequential inroads into the widening of diplomatic relations with the West, the United States blocked any chance to further normalize relations by refusing to recognize the Marxist-controlled government.[273]

The SADF eventually became the major threat to MPLA stability by conducting larger and more sustained military operations in Angola in the late 1970s. South Africa also provided increased support to UNITA, the third revolutionary organization vying for control of the government, enabling it to survive its immediate post-independence defeats and once again evolve into a credible threat to MPLA. Created in 1966 by Jonas Savimbi as a result of his dissatisfaction with the FNLA, the National Union for the Total Independence of Angola

---

[269] John A. Marcum, *The Angolan Revolution, Volume II, Exile Politics and Guerrilla Warfare (1962-1976)* (Cambridge, MA: The MIT Press, 1978), 171-172.

[270] Ibid., 55.

[271] Tom Young, "Angola: Peace at Last," in *South Africa at the Crossroads?*, ed. Larry Benjamin and Christopher Gregory (Rivonia, South Africa: Justified Press, 1992), 24.

[272] Ibid., 26.

[273] Ibid., 29.

(UNITA) became South Africa's main ally, uniting with the SADF to bring the war to a climax in the late 1980s as MPLA forces, with Soviet and Cuban support, attempted to defeat the group in its own territory in southeastern Angola. (For a more in-depth discussion of UNITA, see the section, " South Africa and Its Ally" on page 141.)

The African National Congress. The African National Congress was initially organized to promote the economic and political well-being of black Africans in South Africa. Established in 1912 in Bloemfontein, South Africa, capital of the Orange Free State, the organization was founded on Gandhi's philosophy of nonviolence, its leaders influenced by white missionaries and white liberals in Parliament.[274] After World War II, however, the organization became more militant as white intransigence obstructed rising expectations in the black community. The election of 1948 only exacerbated the problem as the National Party came to power and its leadership immediately initiated a policy of *apartheid,* which in Afrikaans means "apartness or separate."

The South African Communist Party (SACP) soon became a major influence in the ANC. Founded in Cape Town in 1921, the SACP was composed primarily of union workers, both black and white. Armed with a mandate from Moscow, in 1928 the SACP set forth to take control of the ANC when Joseph Stalin directed the party to "explain to the native masses that the black and while workers are not only allies, but are leaders of the revolutionary struggle of the native masses against the white bourgeoisie and British imperialism . . . . Our aim should be to transform the African National Congress into a fighting nationalist revolutionary organization against the white bourgeoisie and the British imperialists, based upon the trade unions, peasant organizations, etc., developing systematically the leadership of the workers and the Communist Party in this organization."[275]

Throughout the 1930s and 1940s the SACP and its Marxist ideals slowly permeated the ANC. A pivotal alliance between the two organizations occurred two years after the 1948 elections when the National Party took a strong anti-Communist stand and banned the SACP.

Although forced to go underground, the SACP remained a potent organization, and its members quickly rose to leadership positions in

---

[274] Allister Sparks, *The Mind of South Africa* (The New Hotfire Trust, 1990; First Ballantine Books edition, New York: Ballantine Books, 1991), 236.

[275] *The Communist International* 4, no. 2 (15 December 1928); quoted in Morgan Norval, *Inside the ANC: The Evolution of a Terrorist Organization* (Washington, DC: Selous Foundation Press, 1990), 33.

the ANC. When the South African government banned the ANC in 1960 following the Sharpeville massacre, SACP members, now ANC leaders, were the most qualified and willing to lead a revolutionary struggle against the establishment and to secure outside aid from the Soviet Union and other Marxist regimes.

In 1961 the ANC established an armed wing, *Umkhonto we Sizwe* (Spear of the Nation), also referred to as "MK." The purpose of this group was to sabotage economic and politically symbolic installations around South Africa. Unfortunately for the ANC, most of its leaders were caught in a police raid in 1963, found guilty of sabotage and other acts of violence, and sentenced to life in prison.

The overt armed struggle resumed in the mid-1970s when a number of confrontations with police, including riots in Soweto during 1976, caused a more militant mood to develop among the ANC's membership. Many left South Africa to join ranks with the exiled MK in training bases located in Angola, Mozambique, and Zambia. Others were sent to camps in Nigeria, Libya, East Germany, Cuba, and the Soviet Union for ideological indoctrination and insurgency training.

Since 1969 a majority of the ANC's military and financial aid from the Soviet Union and its allies, along with rapid growth in the MK, gave the ANC the capability to reinstitute its campaign of violence, and in 1977 the organization set forth on an "armed struggle" against the white South African government. As the struggle intensified through the late 1970s and 1980s, the ANC's commitment to avoid the taking of human life waned. At first only political, military, and economic installations were targeted. Later, key strategic targets such as the Sasolburg oil refinery and Koeberg nuclear power station were added to the list. Soon security force personnel were targeted, followed by farmers, most of whom belonged to the reserve defense force. The distinction became even more blurred as elements of the ANC initiated operations against soft targets where the indiscriminate loss of civilian life became a reality. The turning point appeared to take place in May 1983 when the South African Air Force Headquarters in downtown Pretoria was bombed, resulting in 19 deaths and 217 injuries. Figure 6.3 [SOURCE: Morgan Norval, *Inside the ANC: The Evolution of a Terrorist Organization* (Washington, D.C.: Selous Foundation Press, 1990, 105] illustrates how the terrorist acts increased until ANC activity became a significant threat to South African stability in the mid-1980s.

Although Oliver Tambo, president of the ANC in exile, and other elder members of the ANC disapproved of the change in strategy, the

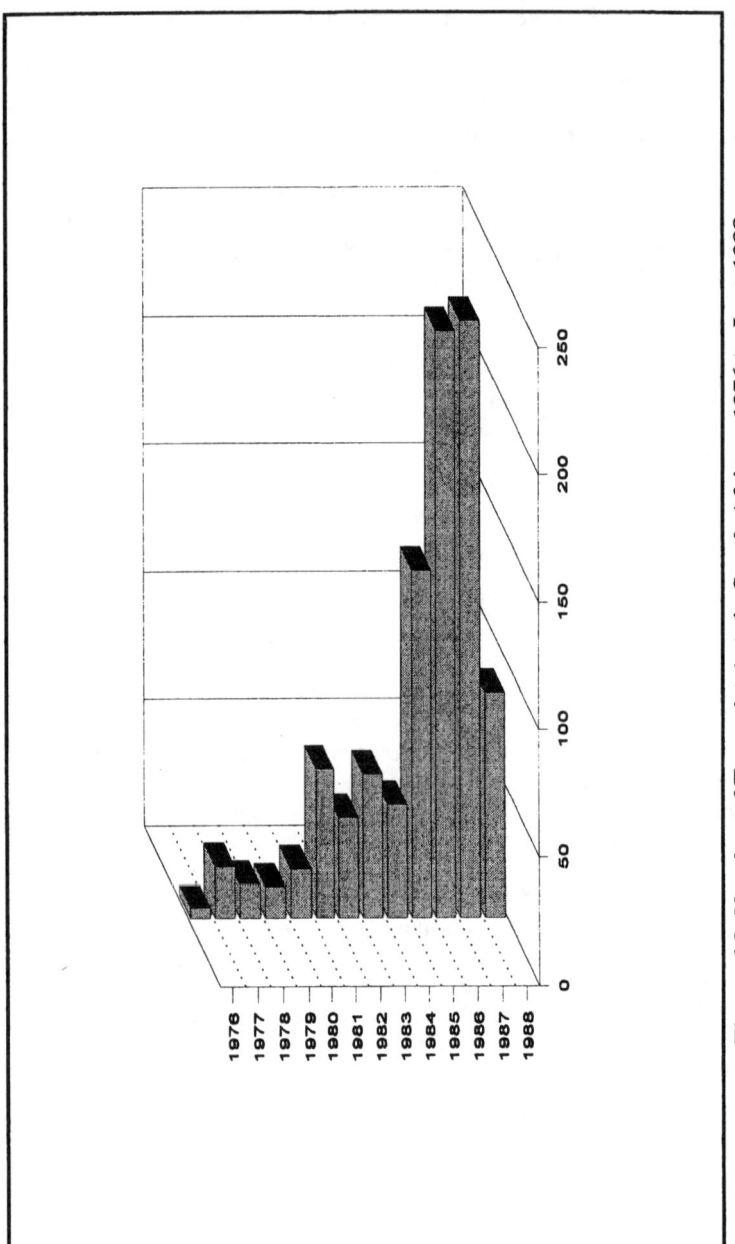

Figure 6.3 Number of Terrorist Acts in South Africa—1976 to June 1988

more militant members of the organization condoned bombing attacks such as the one in Pretoria. One, Joe Slovo, SACP Secretary General and member of the ANC's Political Military Council (PMC), stated that the ANC's shift in strategy was a direct response to "terrorist attacks" by the South African government. In the future, he claimed, MK guerrillas would strike more than just economic targets.[276] Another, Chris Hani, MK chief of staff, said that the whites must be shaken out of their complacency and that part of the ANC's campaign was to do just that: "Their life is good. They go to their cinemas, they go to their braaivleis [barbecues], they go to their five-star hotels. That's why they are supporting the system. It guarantees a happy life for them, a sweet life. Part of our campaign is to prevent that sweet life."[277]

The South West Africa People's Organization. As a result of Germany's defeat at the end of World War I, South West Africa (Namibia)[278] became a League of Nations territory and in 1919 was intrusted to South Africa for administration purposes. In 1945 the trusteeship was transferred to the United Nations, but South Africa, wanting to annex the territory, challenged the status of the new arrangement in the International Court while continuing to administer the territory. This was the first in a series of legal maneuvers concerning SWA's future as an independent country.

In 1960 Ethiopia and Liberia challenged South Africa's right to be in South West Africa and its performance as the administrator in the International Court. Six years later, the UN formally rescinded South Africa's mandate to be in SWA and in 1968 adopted the name of "Namibia" for the country. In 1971 the Court advised against South Africa's subjugation of Namibia. South Africa's refusal to abide by the UN's and Court's directives to grant independence to Namibia

---

[276] *Guardian* (London), 6 June 1983.

[277] *Weekly Mail* (Johannesburg), 16 Oct. 1988.

[278] In 1966 the United Nations formally revoked South Africa's mandate to South West Africa, and it adopted the nationalist name "Namibia" for the country in 1968. Since then the nation has been known by the international community as Namibia. South Africa continued to refer to it as South West Africa until its independence in 1990. Use of each name in this treatise will depend on the context: "South West Africa" will be used when examining the South African perspective, while "Namibia" will be employed in conjunction with a viewpoint other than that held by South Africa.

contributed significantly to South Africa's growing diplomatic isolation in the world community.[279]

The South West Africa People's Organization (SWAPO) was formed in 1960 to liberate the South West African people from South African rule. The organization decided that "political and military efforts in pursuit of national liberation were not contradictory but complementary, and should be pursued concurrently."[280] Following the revocation of South Africa's mandate in Namibia by the UN General Assembly, SWAPO launched its armed struggle with mainly isolated hit-and-run attacks and low-level guerrilla operations against the South African Police (SAP) along Namibia's northeastern border region. Although the level of combat was relatively low, the SAP was soon unable to cope with the fighting, resulting in the first-time, large-scale deployment of South African military forces to the northern part of the country in 1972.

In 1973 the United Nations recognized SWAPO as a "national liberation movement" and declared the organization to be the "authentic representative of the Namibian people."[281] This major change in policy had a profound effect on SWAPO's diplomatic relations. Soviet and Eastern-bloc countries substantially increased aid to the revolutionary group, while African and other Third World countries furnished political support. Members of the People's Liberation Army of Namibia (PLAN), the military wing of SWAPO, were trained by Soviet, Cuban, and East German advisors.

Angola's independence in 1975, along with South Africa's decision to withdraw the SADF from Angola after Operation Savannah, resulted in the establishment of a refuge for SWAPO forces. The bases located in southern Angola provided the guerrilla organization with a safe haven to obtain foreign military support and a sanctuary for cross-border raids into Namibia. As a result, by 1977 there was a significant increase in insurgent activity all along Namibia's northern border. Guerrilla strength in that year was estimated to be more than 3,500, with approximately 300 operating inside Namibia at any given time.[282]

---

[279] Ungar, 298.

[280] SWAPO Department of Information and Publicity, *To Be Born a Nation: The Liberation Struggle for Namibia*, (London: Zed Press, 1981), 176.

[281] Ungar, 298.

[282] Kimmo Kiljunen, "The Ideology of National Liberation," in *Namibia: The Last Colony*, ed. Reginald Green, Marja-Liisa Kiljunen, and Kimmo Kiljunen (Burnt Mill, U.K.: Longman Group Limited, 1981), 158.

The war continued to escalate between 1978 and 1980 as PLAN guerrillas increased the number of sabotage attacks on Namibia's economic infrastructure and military bases, and tried to assassinate moderate black leaders. In 1980 the South African government adopted a more aggressive policy against SWAPO through successive cross-border strikes. These forays, often deep into Angolan territory, forced PLAN forces and their base camps to relocate far north of the border and caused the number of SWAPO guerrillas inside Angola to decrease from 8,000 in 1980 to 6,000 in 1982.[283] Consequently, the volume of insurgent activity in Namibia attributed to SWAPO also decreased. General Meiring, Chief of the South African Army, claimed that the number of guerrillas rose to approximately 9,100 by 1986, but of this total about 3,500 were fighting with FAPLA in Angola against Jonas Savimbi's UNITA.[284]

Despite continued cross-border incursions by the SADF, SWAPO guerrillas were still able to launch deep penetration raids into Namibia. Nonetheless, by 1984 SWAPO did not pose a real threat to South Africa's control over the country. The potential for an open-ended guerrilla war still remained, however, and such a conflict was not attractive to the South African government, particularly since SWAPO was operating increasingly under the umbrella of MPLA and Cuban forces. At a minimum, the SADF would be forced to maintain a continued presence in northern Namibia at a level comparable to the estimated 20,000-25,000 troops they had deployed in 1981.[285]

South Africa and Its Ally, UNITA

While South Africa faced many opponents, it had only one main ally: the National Union for the Total Independence of Angola (UNITA). Several Western nations, including the United States, provided military logistical aid to UNITA on an infrequent basis and several Third World countries occasionally provided the insurgent group with surreptitious assistance and support, depending on the political winds; but it was the Pretoria-UNITA alliance that constituted the major obstacle to the MPLA gaining complete control over Angola.

---

[283] Steenkamp, 106.

[284] John Laffin, *War Annual 2: A Guide to Contemporary Wars and Conflicts* (London: Brassey's Defence Publishers, 1987), 192.

[285] Chester A. Crocker, *South Africa's Defense Posture: Coping with Vulnerability*, The Washington Papers, vol. 9, no. 84 (Beverly Hills, CA: Sage Publications, Inc.), 28.

UNITA. As mentioned previously, UNITA was created as a new opposition movement in 1966 by Jonas Savimbi as a result of his dissatisfaction with the FNLA. It was founded on two main beliefs held by Savimbi: (1) that Angolans from all "tribes, clans, and classes" needed to be involved in the liberation effort (The active groups at the time represented less than half the population.); and (2) that a free Angola should not come from forces outside the country—only Angolans could free the country from foreign rule. Savimbi maintained, with good reason, that the MPLA was a Marxist organization dominated by Moscow, while Roberto's FNLA was "supported by western forces."[286] Despite his opposition to foreign involvement however, Savimbi accepted aid and support from China in the early days of the movement, and later received direct assistance from other African countries, the United States, and South Africa as his fortunes took a turn for the worse.

From 1966 to 1974, UNITA, as well as the FNLA and the MPLA, was relatively ineffective against the Portuguese government in the struggle for independence. However, the combined effect of colonial wars in Angola, Mozambique, and Guinea brought about the 1974 coup in Portugal. The new Portuguese government came to power embracing an anti-colonial theme, which led to Angola's independence and the MPLA's taking control of the government with the help of Cuban and Soviet aid.

In February 1976, with the loss of American resources, the withdrawal of South African troops to Namibia, and unrelenting pressure from Cuban and FAPLA forces supported by Soviet logistics, Savimbi and his UNITA followers began a seven-month retreat to the relative safety of the bush in the savannah of southeastern Angola.

An odyssey of 3,000 kilometers replete with unrelenting combat, exposure, starvation, and exhaustion, Savimbi's "long march" to southern Angola ended in late August 1976. Of the 1,000 followers who began the trek with him, only 79 remained. One of these survivors, Tito Chingungi, later recalled:

> The march was the most profound experience of my life. You felt you needed to love your brother as yourself. Alone, you couldn't survive. When colleagues died you truly felt diminished. All of us who were on the march believed by the end of it that the war really could be won.[287]

---

[286] Marcum, 165-166.

[287] Tito Chingungi, quoted in Bridgland, *Jonas Savimbi*, 218.

Savimbi defied the odds, something he would do time and again. Many emissaries and jounalists concluded that the war had been won with the Cuban-led victory in February 1976. What was not known then was Jonas Savimbi's uncanny ability to survive, especially with South Africa coming to his aid.

The late '70s were lean years for UNITA. The resistance movement received only low levels of covert aid and support from South Africa and several other African nations. This included medical help and combat training for the guerrillas in SWA (Namibia) and a South African government agreement to allow Namibia to be used as a transit point for arms shipments to Savimbi.[288] During this period UNITA was considered as little more than a nuisance to the MPLA rather than as a tangible threat to the stability of the Angolan government.

Bent but not broken, Savimbi targeted the Benguela railway, the main east-west transportation route in central Angola and an important conveyor of Zambian copper and foreign currency producer for Angola. Successive attacks from his guerrilla forces effectively closed rail operations during the late 1970s and produced UNITA's only successful offensive campaign during this time.

In 1979 the SADF substantially increased its military activity in southern Angola. Designed primarily to neutralize SWAPO insurgency in Namibia, this increase was also dedicated to assisting UNITA in its efforts against the MPLA. South Africa's material support was described by Fred Bridgland as a logistical lifeline extending from the Namibian border to 250 miles into Angola. Supplies of arms, fuel, medicines, foodstuffs, and trucks were moved along this route to front-line UNITA forces.[289]

With the SADF's help, Savimbi was able to escalate his campaign significantly, and by the end of 1982 UNITA acquired the capacity to range over all central and southern Angolan provinces. In 1983 and 1984 UNITA was able to launch deep raids into northern Angola, with continued help from the SADF. In addition, anti-Cuban sentiment, a severe economic crisis, and a lack of security were leading to growing support for UNITA among the Angolan people.

Strategically speaking, the possibility of peace talks in southern Africa was frustrated by the difference in objectives between the two warring factions. South Africa felt that the issues surrounding the Angolan civil war and foreign involvement should be settled before addressing the concerns for South West Africa. The Angolans,

---

[288] Jonas Savimbi, interview by Fred Bridgland, London, England, 4-5 July 1980, cited in *Jonas Savimbi*, 258.

[289] *The Sunday Times* (London), 19 July 1981.

Cubans, and Russians, on the other hand, believed the status of SWA should be determined prior to dealing with Angola.

These issues notwithstanding, in 1984 peace negotiations resulted in the Angolan government and Pretoria signing the Lusaka Accords, which provided for a cease-fire in Angola and the withdrawal of South African troops. The SADF began its withdrawal on schedule, but in May 1984 stopped short of completely exiting Angola and established a 40 kilometer-wide buffer zone just north of the Namibian border. In the meantime, FAPLA was being equipped with improved Soviet equipment and receiving extensive organizational and combat training from its Marxist advisors. Ignoring the provisions of the Lusaka treaty, it launched a major offensive in 1985 that threatened UNITA's headquarters in southern Angola.

The South African Defence Force. After the conclusion of Operation Savannah in early 1976, the SADF was permanently stationed in Namibia and defended against a low-intensity insurgency war of cross-border raids by PLAN. By late 1977 it was becoming evident that the insurgency could not be stopped by what was essentially a defensive strategy, where SADF forces reacted only when PLAN chose to raid. In November of that year, Vorster reluctantly agreed to preemptive attacks on SWAPO bases in Angola. This was a pivotal decision: once again South African military force became a significant factor in the Angolan conflict.

Operation Reindeer, initiated in May 1978, called for a South African airborne force to attack SWAPO camps 250 kilometers inside Angola. The first in a series of major SADF external offensives, it was considered a success in that, with the exception of one battle, PLAN kept a low profile in Namibia for the remainder of the year. Activity also remained relatively low the following year with only a few cross-border incursions conducted by the SADF.

South African military operations escalated in 1980, and by 1981 a state of continuous warfare existed in Angola as the SADF attempted to drive SWAPO from the SWA/Angolan border. The long-running, low-intensity conflict in Namibia began spilling over the border into Angola as "hot pursuit" incursions by the SADF evolved into full-scale conventional operations. (See Table 6.2.)

Operation	Date Initiated
Sceptic	May 1980
Shellsmoke	June 1980
Carnation	July 1981
Daisy	November 1981
Protea	August 1981
Super	March 1982
Askari	December 1983

Table 6.2 Major SADF Operations Inside Angola, 1980-1984

SADF leaders maintained that their primary objective in Angola was to eliminate SWAPO's military activity in Namibia. "Our operations in southern Angola are merely a tactic to achieve our aim, which is not to clean up Angola, but to keep South West Africa clean,"[290] contended Acting Officer Commanding of the South West African Territorial Force, Brigadier Willie Meyer, in January 1983.

Launched in December 1983 when a 10,000-strong SADF unit crossed into Angola, Operation Askari became the final offensive of this phase of the war. SADF troops penetrated deep into the country, but Soviet threats to alter the rules of engagement, better resistance from MPLA troops serving under the Cuban umbrella, and unfavorable South African press caused Pretoria to reassess the military situation. It was at this point that Cuba agreed to deploy combat troops in support of the MPLA.

In January 1984, Prime Minister Botha offered to remove SADF troops from Angola if the MPLA and SWAPO pledged that they would not take advantage of the withdrawal. An agreement was reached in February between the contending forces that led to the signing of the Lusaka Accords. This agreement brought a fragile peace to the area and precipitated the SADF withdrawal to southern Angola.

---

[290]*The Star*, 8 January 1983.

South Africa and Angola: 1985-1988

Several events took place in 1985 that contributed to an escalation of the conflict in Angola and propelled the antagonists toward a crucial encounter in 1987. First, in mid-July the Reagan Administration was able to get the Clark Amendment repealed, and for the first time in nine years, the United States was allowed to provide military aid to UNITA. Second, by 1985 UNITA controlled approximately one-third of the land area of Angola with a trained force of 30,000 men.[291] (This territory was economically and strategically unimportant, but gave UNITA a secure base from which to conduct operations.) Third, FAPLA, indifferent to the Lusaka Accords, initiated a new offensive with an expanded and modernized air force, growing numbers of Soviet advisors, and direct support from Cuban soldiers.

The Soviet Union provided Angola with over $1 billion worth of arms between January 1984 and August 1985.[292] Flush with this massive stockpile of modern equipment, FAPLA launched an offensive with the intention of taking UNITA's main logistics base at Mavinga and driving UNITA from its stronghold in southeastern Angola.

The capture of Mavinga would have been a significant victory for FAPLA because the base was vital to Savimbi's insurgency effort throughout Angola. It had an all-weather airfield where critical supplies could have been flown in from South Africa and Zaire. The town also protected the approaches to UNITA's "capital" city of Jamba. The loss of Jamba would have presented a serious psychological blow to Savimbi's revolutionary movement and loosened his hold on the southeastern part of Angola.

UNITA was unable to arrest the FAPLA offensive drive toward Mavinga, therefore, Savimbi was forced to call on South Africa for support. Until this point, South Africa's policy toward Angola since the Lusaka Accords was to refrain from direct intervention in the country. However, with UNITA facing imminent defeat, Pretoria was compelled to intervene with close air support, artillery and medical advisors, air logistics, and artillery support. On 20 September 1985, Defence Minister Magnus Malan said publicly for the first time that South Africa was providing military support to UNITA "of a material, humanitarian, and moral nature." He added that, through its connections with UNITA, South Africa was maintaining the interests of

---

[291] Helmoed-Römer Heitman, *War in Angola, The Final South African Phase* (Gibraltar: Ashanti Publishing Limited, 1990), 13.

[292] Bridgland, *Jonas Savimbi*, 443.

the free world on its subcontinent and that it would only break its links with UNITA on condition that all foreign forces were withdrawn from Angola.[293]

The FAPLA troops were turned back but maintained an important strategic and logistics base at Cuito Cuanavale where they immediately started receiving another massive infusion of military equipment and personnel from their communist allies. Western intelligence agencies estimated that the number of Cuban troops in Angola rose from 25,000 to 31,000 during 1985 and were being supported by 3,250 Soviet and East German personnel.[294]

FAPLA launched a second offensive from Cuito Cuanavale in June 1986, but the assault appeared less intense than previous operations. The objective again was the capture of Mavinga and an attempt to force UNITA from its secure base. This combined force not only consisted of Cuban troops and Soviet advisors, but also included approximately 7,000 SWAPO guerrillas and three 300-man battalions of ANC personnel.[295] UNITA successfully defended its position by employing guerrilla tactics against the invading forces, and eventually brought the advance to a standstill. South African special forces and G-5 (155-mm howitzers) artillery were said to have played a role in UNITA's success, while the introduction of U.S. Stinger anti-aircraft missiles may have prompted FAPLA units to be more cautious this time around.[296]

The two FAPLA offensives in 1985 and 1986, and prospects for more in the future, caused South Africa and UNITA to discuss the possibility of joint planning. Although the relationship was never as congenial as desired during these talks, three objectives for the conduct of defensive action against FAPLA were identified:

- Concentrate on the employment of guerrilla operations.

- Prevent a conventional FAPLA threat to UNITA strategic points during 1987.

- Neutralize FAPLA's ability to launch a conventional offensive by March 1988.[297]

---

[293] Ibid., 446.
[294] Ibid., 442.
[295] Heitman, 17.
[296] Vanneman, 53.
[297] Heitman, 18.

On the MPLA side, the government was under strong pressure to achieve a military victory over UNITA. The Angolan economy was in a shambles because of the war. The MPLA was losing support from people in the urban centers who had backed the movement for over two decades. Foreign governments had poured men and vast quantities of materials into the country for years without any appreciable progress in establishing a stable Marxist regime. Soviet prestige and communist ideology were at stake.

In short, the two Angolan warring factions, along with their allied forces and surrogate armies, were headed down a collision course toward the largest and most intense military action ever encountered in southern Africa. This time the planning and control of FAPLA and Cuban forces would fall under the command of a *Russian* general officer, General Konstantin Shaganovitch.

South Africa Intervenes in Angola

Although FAPLA did not believe it was ready for another strike at Mavinga, Soviet and Cuban advisors felt otherwise. They reminded the revolutionary group that a successful offensive against UNITA in southeastern Angola would:

- Reduce UNITA's guerrilla activity in other parts of the country.
- Restrict UNITA's ability to interfere with the Benguela railway.
- Provide a base to assault Jonas Savimbi's headquarters at Jamba during 1988.
- Lengthen the "safe haven" part of the Angolan/Namibian border to provide more territory for SWAPO to conduct insurgency operations against South African forces in Namibia.
- Allow a modern air defense system directed at the South African Air Force to expand into the southern region.[298]

During the early part of 1987 South African intelligence started picking up indications that FAPLA was preparing for a two-phased offensive: one in the northern part of Angola and one directed toward Mavinga in the south. By April, evidence suggested that a large-scale buildup of forces and material around Cuito Cuanavale would lead to the largest offensive yet against UNITA. As a result, in May South Africa agreed to provide SADF support to UNITA. In June and July FAPLA deployed four infantry brigades and two groups of tank

---

[298] Ibid., 21-22.

battalions to the east of the Cuito River. It held four battalions in reserve in Cuito Cuanavale.

The SADF leadership presented several options to the South African government in order to counter the expected FAPLA offensive. The primary plan included a major attack on Cuito Cuanavale from the west to disrupt the logistics flow to FAPLA troops. The South African Cabinet vetoed this option and instead approved the secondary plan: to set up a defensive force east of the Cuito River to prevent Mavinga from being captured. This seemed the better choice to State President P.W. Botha and his Cabinet, who were acutely concerned about the international outcry should a South African force be found fighting deep inside Angolan territory and also about the national response should a significant number of white soldiers die under these circumstances.

Consequently, in approving the SADF plan, the government directed that the FAPLA offensive must be halted, but that the SADF role would be defensive only, and that any success would be attributed to UNITA. According to Bridgland, the SADF's involvement was to remain secret and sufficiently limited to be "plausibly deniable." "No men must be lost; no equipment must be lost; and you must achieve all your objectives."[299] This political decision to fight with minimal force in what would soon become a conventional war would cause the SADF to forfeit many opportunities to completely destroy the FAPLA over the coming year.

Operation Modular

Operation Modular, the first of three South African operations in the war, was carried out under the command of the Chief of the Army, Lieutenant General "Kat" Liebenberg. It was conducted in four escalating phases:

1. Watch and harass the FAPLA as it deploys.
2. Monitor and harass the FAPLA as it advances.
3. Stop the FAPLA advance.
4. Destroy the FAPLA units involved.[300]

In mid-July 1987, the SADF moved 32 Battalion into Angola to counter the combined 10,000-strong FAPLA and Cuban force that was slowly advancing toward Mavinga. It was obvious from the beginning that the size of the force employed and the rules of engagement (ROE)

---

[299] Bridgland, *The War for Africa*, 33.
[300] Heitman, 31.

were inadequate for the stated mission. The deputy officer commanding the South African forces in eastern Angola, Commandant Jan Hougaard, summed it up this way:

> We were really only monitoring the situation. I think at certain times we were confused by the orders we got *not* to get into a fight; only use artillery when we were really in trouble or UNITA was in big trouble.... Our orders were *not* to stop the advance but to prevent the enemy from taking Mavinga.[301]

The original operational concept for South African forces was to make an initial stand at the Chambinga River, and then to break contact and conduct continual harassing operations against the main FAPLA troops advancing toward Mavinga. Harassment operations would also be conducted in FAPLA's rear areas to hinder their logistics flow.[302]

As 32 Battalion began its deployment into Angola, the Chief of the Army and Chief of Staff for Operations immediately saw that the authorized force was too small for the defined mission, so they approved the deployment of an anti-tank squadron (Ratel-90s) and a battery of G-5 long range artillery. They also initiated planning for the deployment of 61 Mechanised Battalion, which consisted of a whole range of armored vehicles (not including tanks). Early restrictions, however, dictated that these additional forces would be used only as a last resort in a final defensive battle.[303]

By 17 August four FAPLA offensive brigades (16, 21, 47, and 59) had deployed across the Cuito River and were advancing on Mavinga. A high-level meeting at Rundu, a base camp in northern Namibia now serving as brigade headquarters for South Africa's forces in Angola, was held on 28 and 29 August with Generals Geldenhuys, Liebenberg, and Earp (Chief of the South African Air Force). Figure 6.4 illustrates the area of operations.

---

[301] Bridgland, *The War for Africa*, 36.
[302] Heitman, 34.
[303] Ibid., 39.

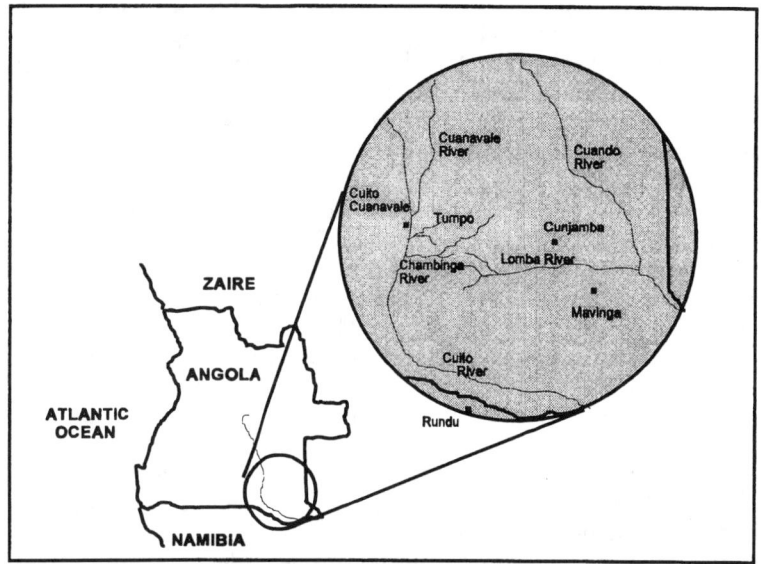

Figure 6.4 Area of Operations

The criticality of the FAPLA advance dictated that the generals view the situation in realistic terms. A decision was made to lift the restrictions on 61 Mech, and the battalion was ordered to move immediately to the front and engage the FAPLA forces. An ad hoc 20 Brigade was established to command all South African forces in the theater. Finally, restrictions were eased on the air force so that it could provide offensive air operations in support of the ground troops at the discretion of General Earp.

The meeting of the chiefs of staff in the "forward area" was a *modus operandi* that would hamper commanders in the field throughout the war. Charged with containing the FAPLA offensive until South African reinforcements arrived, then commander of SADF ground troops in Angola Colonel Harris stated his position succinctly: "There was a very fast and furious 10 days, a very panicky period that was. Everyone [general officers] was flown in to bloody well advise."[304] Even with the FAPLA troops closing in on Mavinga, Harris was prevented from using the G-5s because the group at Rundu had to ensure the target was worthy of using those guns.[305]

---

[304] Ibid., 44.
[305] Ibid., 45.

Initial South African Air Force operations were restricted to cargo flights between Rundu and Mavinga. Impala and Mirage strike aircraft were deployed to northern Namibia in early September, but the first offensive air strike did not occur until 16 September 1987. As with the ground operations, once air strikes began, many restrictions were placed on offensive air operations by SADF headquarters in Pretoria, mainly in response to political considerations.

Although the first contact between FAPLA troops and South African ground troops occurred on 6 September 1987, the first major battle did not take place until 10 September when 32 Battalion attacked a part of FAPLA's 21 Brigade as it attempted to cross the Lomba River, approximately 35 km from Mavinga. The river crossing was halted by the SADF attack; however, 21 Brigade continued to amass supplies expecting to attempt another crossing as soon as practical. The brigade was finally forced to abandon the effort on 27 September due to an unrelenting, accurate barrage from the G-5 artillery that was at last allowed to join the fight. When the decision was made to order 21 Brigade to pull back from the Lomba, the unit was reported to be only one-third of its original strength.[306]

Impacting the results of this first engagement, the G-5 artillery soon became South Africa's most effective weapon. SADF reconnaissance teams would repeatedly penetrate FAPLA lines, pinpoint military positions, and then call in highly accurate artillery fire. This tactic minimized exposure to South African forces, which accounted for the small number of SADF casualties with respect to the number of killed and wounded FAPLA combatants.

Political considerations also frustrated military operations when SADF troops engaged FAPLA's 47 Brigade as it advanced toward the Lomba River in early September. Throughout the brigade's advance the unit was subjected to a constant artillery barrage, and was finally brought to a standstill at the river where its position was marked for an air strike on 11 September. Unfortunately for the South Africans, the mission was called off at the last moment because negotiations were entering a critical stage for the release of a South African officer who had been captured in Angola in 1985.

Nevertheless, G-5 shelling continued through the month, and in early October the 47 Brigade was ordered to withdraw from its position near the river. In a fast-moving battle on 3 October between the retreating brigade and SADF 61 Mech, the FAPLA unit was defeated so completely that it was no longer considered to be an effective fighting force and remained in that condition for the duration of the

---

[306] Bridgland, *The War for Africa*, 101.

campaign. For all intents and purposes, the entire FAPLA offensive was stopped with this one battle, which essentially accomplished the objective of the first phase of Operation Modular.[307]

Even before the 47 Brigade's defeat, the operation appeared to be turning in favor of SADF/UNITA forces. A meeting was therefore convened on 28 and 29 September at Brigade Headquarters that included: President P.W. Botha; the Minister of Defence Magnus Malan; the Deputy Minister of Defence Waynand Breytenbach; General Geldenhuys; General Liebenberg; the Chief of Staff Intelligence Admiral Putter; and the Chief of Staff Finance General Willie Meyer. In the course of the meeting, this group of high-powered leaders decided to take a more offensive posture by adding the objective of attacking and destroying all FAPLA troops deployed east of the Cuito River before they could regroup and launch another assault on Mavinga. The intent was to defeat the FAPLA troops so badly that they would not be able to launch an offensive in the following year. President Botha assured General Geldenhuys that the necessary additional forces and funding would be made available to support this operation.[308]

The SADF could not follow up on initial successes against 21 and 47 Brigades immediately, however, because it lacked adequate ground forces and did not possess battle tanks. In addition, UNITA was not trained in conventional warfare, and the logistics system could not be expanded quickly enough.[309] Jan Hougaard complained:

All our planning had been for a defensive operation. Overnight we were meant to switch onto the offensive and begin a long-distance chase of a conventional army. Colonel Ferreira [Brigade Commander and Officer Commanding of all South African troops in Angola] foresaw that as we moved north our logistics problems would multiply. We already had very, very long logistical lines. It was incredible: you're not actually meant to wage war in that way, army academies tell you.[310]

This delay allowed the defeated FAPLA to withdraw in good order and to regroup the shattered remains of the two brigades.

Operation Modular's new mission was defined on 6 October. The SADF was to work in concert with UNITA forces to accomplish a three-phase operation:

---

[307] Ibid., 163.
[308] Heitman, 71-72.
[309] Ibid., 79.
[310] Jan Hougaard, quoted in Bridgland, *The War for Africa*, 172.

Phase 1 - Stop the FAPLA offensive (essentially completed by 3 October).

Phase 2 - 20 Brigade to gain the initiative by 6 November, using only existing forces, and inflict maximum losses on FAPLA units east of the Cuito River.

Phase 3 - 20 Brigade and UNITA forces to destroy the FAPLA troops east of the Cuito River by 15 December.[311]

South African forces deployed to initiate the second phase of Operation Modular on 11 October, but operations were slow to materialize because of a breakdown in the logistics system. A number of issues came into play:

- The distance between Mavinga and the forward forces in the Mianei River area was large.
- The demand for more supplies and equipment was greater than planned.
- There were not enough vehicles to support the logistics operations.
- Artillery ammunition could not be stockpiled because of the mobile nature of operations.
- Air transport from Rundu to Mavinga was restricted to night operations, which caused a loss of efficiency in the system.[312]

During the last half of October the SADF received some much needed reinforcements organized under a new group called Combat Group Charlie. Three G-6s, self-propelled versions of the G-5, were brought on-line. (In the prototype stage of development, these guns were still undergoing testing, and tooling for production lines had not been completed.) The G-6s were joined by a squadron of 13 Olifant tanks (the SADF's main battle tank) and a second battery of eight G-5 guns.

The convoy of reinforcements arrived in Mavinga on 30 October, joining up with a second group, Combat Group Alpha. Even with the arrival of this additional force, the South African contingent in Angola amounted to only 3,000 men, while the FAPLA and Cuban units with

---

[311] Heitman, 84.
[312] Ibid., 102.

their Soviet advisors in and around Cuito Cuanavale now totaled more than 15,000 soldiers.[313]

The difference was made up in the air. Restrictions on the SAAF were eased on 16 October. Although aircraft still could not be used to attack infrastructure targets, the Impala and Mirage jets were now allowed to better support the ground forces. As for FAPLA, its air base at Cuito Cuanavale was no longer in regular use due to continual shelling from the G-5 batteries. This forced its air operations to originate at the logistics base at Menongue, thereby cutting Mig operational effectiveness considerably due to the increased range to target. The threat from Mig air-to-ground bombing operations remained a concern for SADF commanders for the remainder of the war, but the Migs were no longer capable of seriously influencing the outcome of a battle.

By the end of October the SADF considered Phases 1 and 2 complete. A new Operations Instruction (OI) was issued by General Liebenberg for Phase 3. It included the destruction of all FAPLA brigades east of the Cuito River and identified a possible Phase 4 with the objective of seizing Cuito Cuanavale should the opportunity arise. A likely impediment to these operations was that all national servicemen involved in the fighting were scheduled to be demobilized at the end of 1987, which would require them to be released from the Angolan theater of operations by 15 December.

Meanwhile, FAPLA troops had suffered serious casualties in the fighting, and by 1 November approximately 15,500 out of an original 18,000 complement remained.[314] (The discrepancy in the quoted numbers of FAPLA troops must be attributed to the fact that the numbers were only estimates.) Castro considered a defeat of the MPLA government as a defeat for the Cuban Revolution since Cuban troops were involved in the war in significant numbers. He therefore ordered in additional troops, aircraft, pilots, and several of his best generals. By 15 November 15,000 Cuban reinforcements had arrived in Angola, and Cuban generals began taking over command of the follow-on counteroffensive. FAPLA commanders were being forced to assume a subordinate role.[315]

The third phase of the operation began on 9 November with an attack by a combined SADF/UNITA force led by Combat Group Charlie on FAPLA's 16 Brigade, now located at the source of the Chambinga River. This was the strongest fighting force yet put on the

---

[313] Bridgland, *The War for Africa*, 183.

[314] Heitman, 118.

[315] Bridgland, *The War for Africa*, 342.

field by the SADF in more than a decade of South African military involvement in Angola. Nevertheless, the SADF was still significantly weaker than its opponents in terms of numbers and firepower, even with the addition of the Olifant squadron.[316]

The first day of fighting was inconclusive although 16 Brigade sustained heavy losses and was forced to withdraw as its defensive line was breached by the Olifants. But the SADF was unable to profit from its advantage as the "fog of war" hindered repeated attempts to fully overwhelm the brigade. As a result, Combat Group Charlie had to fall back to rest and regroup while 16 Brigade was allowed to retreat. The loss of men and material was one-sided, but the attack on 16 Brigade had to be considered a failure since the main objective of eliminating it as a viable fighting force was not realized.

On 11 November Combat Groups Charlie and Alpha again attacked FAPLA's 16 Brigade with the same results: the brigade took devastating losses but was again able to withdraw without being totally defeated. The failure to destroy 16 Brigade also left an opening for FAPLA's 66 and 59 Brigades to retreat to safety in Cuito Cuanavale.

Around mid-November, G-5 batteries began suffering from severe stress, placing increased importance on the remaining guns and equipment. From this point on, weapons fatigue and logistics problems played a significant role in the SADF's ability to sustain combat operations.

On 18 November, Generals Geldenhuys and Liebenberg and Admiral Putter discussed the future of Operation Modular. Three options appeared open to them:

1. Regard Operation Modular as complete and return all SADF troops to South Africa.
2. Drive the FAPLA back across the Cuito River.
3. Take Cuito Cuanavale from the west and destroy all FAPLA units in the theater of operations.[317]

After discussions with the Tactical Headquarters staff, General Liebenberg decided that the FAPLA units east of the Cuito River should be attacked and destroyed or driven back across the river into Cuito Cuanavale. The purpose of this phase was to " systematically grind down and break up FAPLA east of the Cuito River, as well as

---

[316] Ibid., 183.
[317] Heitman, 156.

destroying or driving it off if possible, and dominating the Chambinga bridge and high ground."[318]

The proposed attack was contrary to restrictions placed on operations. When the original decision was made to pursue FAPLA as its units withdrew from the Lomba, the instructions were to destroy FAPLA troops but not to authorize operations north of the Chambinga River. The decision to expand these operations was made "in light of political developments," as Cuba and Angola were making "diplomatic noises" concerning a negotiated settlement to the war.[319] The South African Foreign Affairs Department believed that if the attack succeeded, enough pressure would be placed on the Cubans and MPLA to join in meaningful negotiations.

The last battle associated with Operation Modular occurred on 25 and 26 November when SADF and UNITA forces attacked 66 and 25 Brigades near the Chambinga Bridge. The thick bush, a lack of good intelligence, and heavy fire from FAPLA artillery located in Cuito Cuanavale frustrated the SADF offensive, and the last two days of battle were inconclusive.

Most of the South African men fighting in Angola were national servicemen who were coming to the end of their two-year commitment. Military logic would dictate that their period of duty be extended to finish the campaign while the SADF had the momentum, but internal political considerations mandated otherwise. The South African public had been poorly informed about SADF involvement in Angola: "Disinformation, secrecy and censorship had combined effectively to hide the true scale of the conflict from the Republic's citizenry."[320] Any extension of the national servicemen's tour in Angola would have certainly undermined the government's program of duplicity against its own citizens.

The military commanders in the field resented the "political" decision to replace the battle-tested national servicemen. The replacement group of soldiers would have completed one year of training, but not in "bush" warfare. Additional training would delay further offensive operations and would allow the FAPLA units invaluable time to regroup, rearm, and construct an improved defensive system on the eastern side of the Cuito River. Nonetheless, the decision was made to continue the campaign, and Operation Hooper was born.

---

[318] Ibid., 161.
[319] Ibid., 162-163.
[320] Bridgland, *The War for Africa*, 225.

Operation Hooper

Operation Modular succeeded in stopping the FAPLA offensive to take Mavinga; however, the SADF did not destroy the major war fighting capability of the army, and FAPLA still remained in control of Cuito Cuanavale and held a bridgehead on the eastern side of the Cuito River.

Once the decision was made to remain in Angola and continue operations against FAPLA units around Cuito Cuanavale, General Liebenberg ordered troops to destroy the bridge across the Cuito River and defeat FAPLA units trapped on the river's east side. It was decided not to attack and occupy Cuito Cuanavale itself for the following reasons:

- Additional international pressure would be placed on the South African government for holding an Angolan city.

- Once occupied, the small SADF/UNITA force would find it difficult to defend a stationary target from a larger, better equipped force.

- Should the SADF occupy the town and then be forced to withdraw, the operation would play into the hands of the MPLA propaganda machine.[321]

Setting 31 December 1987 as the date for destruction of these FAPLA units, the SADF identified the objectives for Operation Hooper as follows:

- Exert pressure on FAPLA troops on both the east and west sides of the Cuito.

- Disrupt the logistic flow of supplies to Cuito Cuanavale and the FAPLA units east of the river.

- Capitalize on every opportunity to destroy the FAPLA.[322]

The month of December was conspicuous, however, for its lack of military action. FAPLA concentrated on fortifying positions on the east side of the Cuito. The SADF attempted to disrupt resupply efforts

---

[321] Heitman, 169.
[322] Ibid., 178.

across the Cuito River bridge with artillery and several "prototype" smart bombs, but to no avail until 3 January when a smart bomb released from a Mirage dropped a portion of the span. Meanwhile, new SADF national servicemen were being trained in the art of bush warfare, and planning for the next operation of the war had to be finalized. These extensive preparations forced General Liebenberg to roll back the operational start date of Hooper until early January 1988.

The first major military ground action of Operation Hooper took place on 2 January when UNITA forces conducted an ineffective attack on FAPLA's 21 Brigade. Failing to move the brigade's defensive positions, this feeble attempt helped to confirm South African convictions that UNITA, lacking extensive training in conventional warfare, was not capable of successfully defending itself without heavy support from the SADF.

A second attack was launched against 21 Brigade on 13 January by a combined SADF/UNITA force. The brigade was forced to retreat from its defensive position, but, as in the past, the South African force was unable to capitalize on its initial success for the same reasons that had hampered its operations throughout the entire campaign, too small a force and logistics limitations. Even with these limitations some units of 4 SAI (South African Infantry) actually penetrated to within two kilometers of the bridgehead before they were "ordered to withdraw for political reasons" by higher headquarters in Pretoria.[323] FAPLA once again had time to regroup and eventually reoccupied the same defensive positions it had been driven from earlier in the campaign.

Throughout the rest of January and most of February the SADF, with help from UNITA, scored modest successes, but remained unable to inflict a decisive defeat. In the end, FAPLA remained in control of the bridgehead on the east bank.

Even at this late stage of the war, battlefield commanders were still being hampered by the micro-management style of the South African government. For his intended attack on Tumpo in mid-February, Commandant Mike Muller, the Officer Commanding of 61 Mech, was required to send his battle plan for the attack through the Chief of the Army to the Minister of Defence for approval.[324] In defense of the government at this point, however, important political and diplomatic activities were occurring on the international scene that could have been significantly affected by the success or failure of military operations in southeastern Angola.

---

[323] Colonel Jan P. Malan, interview by author, 30 August 1993, Upington, South Africa.

[324] Bridgland, *The War for Africa*, 295.

Further SADF operations in early March were not possible because of the need to again relieve national servicemen who were finishing their tours and to redress efforts to improve the logistics situation. As these problems continued to delay military operations, General Meyer became concerned that events taking place in the international arena might force the South Africans to pull out of Angola before they had successfully completed their mission.[325]

Operation Packer

Operation Packer was initiated on 23 March 1988 with a third attack on the Tumpo region. The attacking force included the newly deployed 82 SA Brigade and four UNITA battalions; the defending units consisted of FAPLA's 25 and 66 Brigades. The SADF objective was to destroy the FAPLA units or drive them from the east bank of the Cuito river and then withdraw.

As with the first two, the assault failed as the result of a slow advance caused by extensive land mines dispersed along the avenue of attack and withering fire from artillery located on the high ground across the river. Although the South Africans did not sustain any casualties, they had three tanks damaged from mines and were forced to discontinue the attack.

It was now apparent that a much larger force would be required to forcibly remove FAPLA from the east bank of the river, a force the government in Pretoria was unwilling to deploy. Given this and the fact that diplomatic initiatives were beginning to overtake military action as the determining factor in shaping the future of southern Africa, the decision was made to discontinue the campaign and bring Operation Packer to a close.

The South Africans did not, however, immediately abandon the ground they had recently expended so much effort and so many resources to win. SADF units remained deployed in the area while minefields and other obstacles were constructed to impede any further FAPLA attempt to advance toward Mavinga. Other SADF troops trained UNITA forces in the art of conventional warfare. Meanwhile, artillery was used sporadically to maintain pressure on the FAPLA, and deception was used to make it appear that a large SADF unit remained in the area.

In April Operation Packer merged into Operation Displace, which evolved into a mechanism for South African troops to withdraw from Angola as negotiations continued in another part of the world.

---

[325] Heitman, 265.

Operation Displace included laying mines, training UNITA forces, and conducting deception activities during the several-month withdrawal. The last South African soldier was finally redeployed to South West Africa on 29 August 1988, just two days short of a year after the SADF entered Angola to blunt the initial FAPLA offensive against Mavinga.

The War in the West

When the Lusaka Accords were signed in 1984, Cuban troops in western Angola redeployed to the interior of the country and remained approximately 300 kilometers north of the South West African border. This situation was the status quo until late 1987 when the FAPLA offensive against Mavinga stalled against the SADF. With the possibility of military failure looming in southeastern Angola, Fidel Castro decided to look west in an attempt to enhance Cuba's strategic position should negotiations bring the war to an end. "Presented as a thrust to compel the SADF's withdrawal from Angola, the potent Cuban force was primarily a political demonstration in keeping with Castro's 'strutting cock' school of grand strategy. There was minimal risk in 'occupying' a large empty place,"[326] wrote Chester Crocker, describing Castro's move to the west.

By January 1988 approximately 3,500 troops of Cuba's 50th Division had deployed into southwestern Angola, with the number increasing to 11,000 infantry and 105 battle tanks by May.[327] Cuban bases were established as far south as Ongiva, just north of the South West African border, and Cuban troops soon became involved in countering SADF cross-border raids against SWAPO.

Negotiations seemed to be progressing, and South Africa correctly assessed the Cuban southward deployment as primarily a "face saving" move; nevertheless, Pretoria decided it could no longer ignore the buildup along the border, particularly since the Cuban advance was approaching the large Calueque-Ruacana hydropower and water project that provided electricity and water to northern Namibia. The South African government therefore sent a sizable force into Angola to make contact with the Cubans around Techipa, approximately 70 kilometers from Calueque. As usual, resources provided for the mission were inadequate, and the campaign turned out to be little more than a harassment action. The last battle between SADF and Cuban troops of

---

[326] Chester A. Crocker, *High Noon in Southern Africa: Making Peace in a Rough Neighborhood* (New York: W.W. Norton & Company, 1992), 371.

[327] Steenkamp, 161.

any consequence took place from 26 to 27 June 1988 and had little military significance.

On 27 June, Pretoria ordered the withdrawal of all SADF troops from western Angola. The conclusion to the decade-long war would be determined at the negotiating table.

A Negotiated Settlement

History has demonstrated that negotiating the end to war is no easy task. The negotiation process that resolved the Angolan War in a manner acceptable to all concerned parties proved to be no exception. The following is a profile of events leading up to the final negotiated settlement:

November 1987. Cuba called on South Africa's Mission to the United Nations in New York to explore a possible negotiated end to the war.

Late January 1988. In a meeting between Chester Crocker, Angolan Foreign Minister Alfonso Van Dunem Mbinda, and Cuban Politburo member Jorge Risquet in Luanda, Angola and Cuba accepted the requirement that the 50,000 Cuban troops would have to be withdrawn as part of any negotiated settlement.

Late January 1988. West German politician Franz Josef Strauss, intermediary for the Soviet Union, told South African Foreign Minister Pik Botha that the Soviet Union wanted to find a political settlement to the Angolan problem because a military solution was no longer possible.

Early February 1988. At a meeting in the South African Embassy in Geneva, Switzerland, Crocker told Botha and a high-ranking SADF officer that the Cubans were ready to negotiate.

March 1988. A letter was delivered to Botha from Crocker proposing talks in London between the Angolan government, the Cubans, and South Africa, with the United States facilitating.

3-4 May 1988. The first official round of talks was held in London between the three adversaries. Crocker chaired the meeting while the Soviet Union's Deputy Foreign Minister Anatoly Adamishin waited in the wings to help if needed.

29 May to 2 June 1988. At their fourth superpower meeting in Moscow, U.S. President Ronald Reagan and Soviet leader Mikhail Gorbachev set 29 September, the tenth anniversary of UN Resolution 435 on Namibian independence, as the target date to settle the Angolan conflict.

23-25 June 1988. A second round of meetings was held in Cairo, Egypt. The Soviet Union forced the Angolan/Cuban delegation to negotiate in realistic terms.

11-13 July 1988. On an island in New York harbor, South Africa, Angola, and Cuba agreed on 14 principles that would bring the war in Angola to an end. These came to be known as the "New York Accords."

5-8 August 1988. Talks in Geneva set detailed conditions for the final agreement, including the total withdrawal of all South African forces from Angola no later than 1 September 1988.

13 October 1988. An agreement was reached on the timetable for Cuban troops to be withdrawn from Angola.

22 December 1988. Botha, Van Dunem Mbinda, and Cuban Foreign Minister Isidor Malmierca Peoli signed the trilateral agreement at UN Headquarters in New York that formally brought the war to an end.

Many factors came together to ensure the success of the negotiations. Within southern Africa, fatigue, lack of purpose, cost, public awareness, other options, and U.S. presence contributed to the settlement. South African forces had stopped the FAPLA/Cuban drive on Mavinga. Now, combat-weary, they could not hope to achieve anything more by continuing a low-level conflict or escalating the fighting to a higher threshold. Meanwhile, the war was estimated to be costing South Africa, whose economy was already in a recession, over $1 million a day.[328] In addition, even though Pretoria maintained tight control of the press, casualties suffered in Angola were slowly becoming known to the public, a dangerous prospect given the government's evasiveness. In addition, it appeared that the MPLA would not have the resources necessary to defeat Savimbi, and South Africa would not allow a UNITA loss should the war continue. Finally, the United States was now in a position, after reinitiating vital logistical support, to pressure Savimbi into the peace talks.

The major changes sparking successful negotiations, however, came as a result of events external to the African continent. For example, Soviet and Cuban men and logistical support for Luanda were placing a drain on these communist countries' already moribund economies. In addition, the Soviet Union as a world power was on the verge of collapse and needed to pull back from its foreign intrigues, Afghanistan being the most prominent example at the time. A new world order was

---

[328] Berridge, 465.

slowly taking shape too, and for the first time the United States and Soviet Union joined together in forging a peace settlement.

Although the war itself was long and frustrating, and the soldiers who fought came away disillusioned about the whole affair, war termination resulting from the negotiated peace accords was a win-win situation for all the parties involved. Namibia gained its independence and conducted free elections in 1990. The war in Angola was finally brought to a close and free elections were held in 1992. (Unfortunately, Savimbi made a grab for power after the elections and reignited a civil war that continues today.)

For South Africa it was a watershed event. The external communist threat disappeared as Soviet influence waned in all countries along South Africa's northern border. The SADF no longer had to counter insurgent threats outside South Africa. And the ANC was forced to abandon its bases in Angola. In the end, the southern Africa region experienced a stability, at least temporarily, unlike any in the area since European colonial powers were in control 30 years earlier.

Chapter 7

Application of the Principles of War

Analysis of the principles of war in the Persian Gulf and Angola provides a perspective on the role each principle played in the decision-making process and the conduct of each war. Additionally, an analysis of both cases reveals a correlation between the appropriate use of principles and the favorable outcome of war.

Unless stated otherwise, observations should be considered applicable at the strategic and operational levels of war. Also, unless specifically cited, quotations used in this chapter were extracted from comments written in the "Principles of War" Questionnaire found in the appendix. Developed by the authors, this survey was used to provide empirical data based on the principles of war to compare the two cases.

A synopsis follows of how well the 40 respondents to this questionnaire believed each principle of war was applied. Although the ratings are based on a numerical scale of one to 10 (with 10 being the highest), in the end the absolute values are not as important as the relative relationship of the ratings between the Gulf War and the Angolan War. Discussions referring to responses from survey questionnaires are based on the most prevalent beliefs expressed in the raters' comments.

The Gulf War Case

Respondents to the survey rated use of the principles of war in the Persian Gulf particularly high, which is not surprising considering the success of military operations there. Figure 7.1 displays the mean and standard deviation of responses to the Gulf War survey.

170   *Applications of the Principles of War*

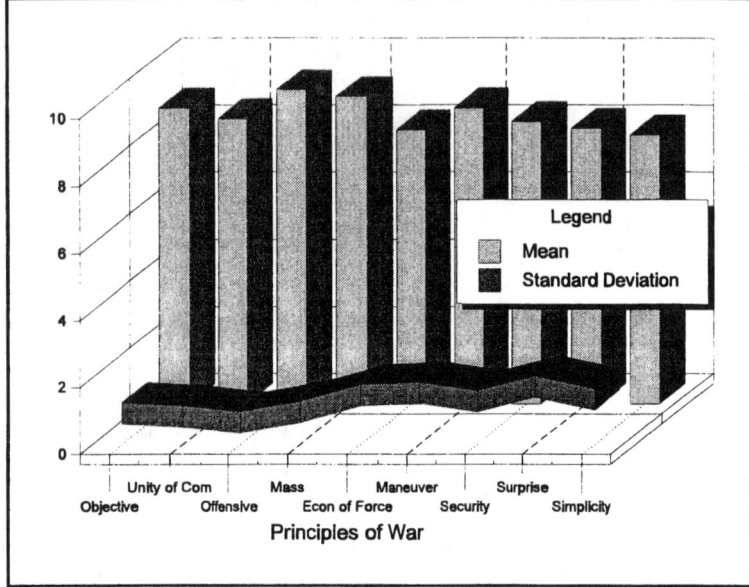

Figure 7.1 Gulf War Survey Mean and Standard Deviation

- **Objective**   <u>Mean</u> 8.8

President Bush announced the political objectives of the Gulf War within days of Iraq's move into Kuwait: the immediate, unconditional, and complete withdrawal of Iraqi forces from Kuwait; the restoration of Kuwait's legitimate government; the security and stability of Saudi Arabia and the Persian Gulf; and the safety and protection of the lives of American citizens abroad. These objectives remained the same throughout the crisis and provided the guidance for military operational planning. Every step of the planning process for the anticipated offensive was based on how to accomplish these original objectives, right down to selection of the targets.[329]

---

[329] Some critics have stated that the overthrow of Saddam Hussein should have been added to the list of objectives. If this were the case, the added requirement to invade Baghdad would have led to a large number of Coalition casualties, while achieving the objective would have been problematic at best. In addition, deliberately targeting Hussein himself would have gone against the American values of fair play and the proper way of conducting warfare.

Results from the questionnaire revealed that the respondents had a suitable understanding of the declared objectives. One additional objective often cited, but never directly declared as a purpose for intervention, was the desire to maintain the oil flow from the gulf. Many believed this to have been an objective that was never declared publicly for political reasons.

- **Unity of Command** <u>Mean</u> 8.5

General Schwarzkopf recently wrote:

> Officially, as a commander in chief, I reported to Secretary Cheney, but Colin Powell was virtually my sole point of contact with the Administration. "It's my job to keep the President and White House and the secretary of defense informed," Powell would say. "You worry about your theater and let me worry about Washington." This arrangement was efficient: I'd tell Powell we needed to get something done in Washington and he'd make sure it happened.... Not since General George Marshall during World War II had a military officer enjoyed such direct access to White House inner circles, not to mention the confidence of the President.[330]

Schwarzkopf's statement lends a great deal of credence to the changes made in the chain of command as a result of the Defense Reorganization Act of 1986. The Chairman was placed in his proper role as a conduit between the CINC and the civilian leadership. Military advice to the President and Secretary of Defense was coming from one source. In short, the civil/military decision-making process worked as designed. According to General Powell:

> [W]e were blessed with a group of political leaders, a President, and a secretary of defense who ... allowed the military to participate in the decision-making process from the very beginning, and allowed me as chairman to be a part of the inner sanctum.... There was, as close as possible, integration between political issues and political thinking and military issues and military decisions.[331]

Powell later echoed Schwarzkopf's praise of the Defense Reorganization Act by stating that it enhanced his position as the President's military

---

[330] H. Norman Schwarzkopf, *It Doesn't Take a Hero* (New York: Bantam Books, Original Paperback, 1993), 377-378.

[331] *Triumph without Victory*, 95.

advisor. He was able to advise based on his own convictions, which was not possible in the past.[332]

At the operational level, placing all Western forces under the control of General Schwarzkopf established unity of command for these forces. Placing all Arab forces under Saudi command (which created a dual chain of command), however, had the potential for failure. Fortunately, the C^3IC was used successfully to ensure coordination between the two chains of command.

Survey responses focused mainly on the U.S. chain of command as presented in Figure 7.2. This arrangement provided the CINC with overall command and control of all U.S. forces in his AOR. Results of the survey indicated that this arrangement worked very well during the war.

- **Offensive**   Mean 9.4

The principle of the offensive was the highest rated principle of war in the Gulf War survey, reflecting the results of the highly successful military operations during the war. One response summed up the elements that made the offensive operations such a success: "Air supremacy; technology; Intelligence (Sattelite [sic], JSTARS, etc.); superior weaponry; well trained soldiers and airmen; flat terrain; Iraqi mistakes; a good plan that was well executed."

The ability to carry out such a well-planned and -executed operation was based on clear objectives, the chain of command, few political constraints, civilian confidence in the military, and an overpowering military force. Appropriate decision making at the civilian/military seam was critical to the military achievement.

---

[332]General Colin Powell, interview by author, 10 May 1994, Pretoria, South Africa.

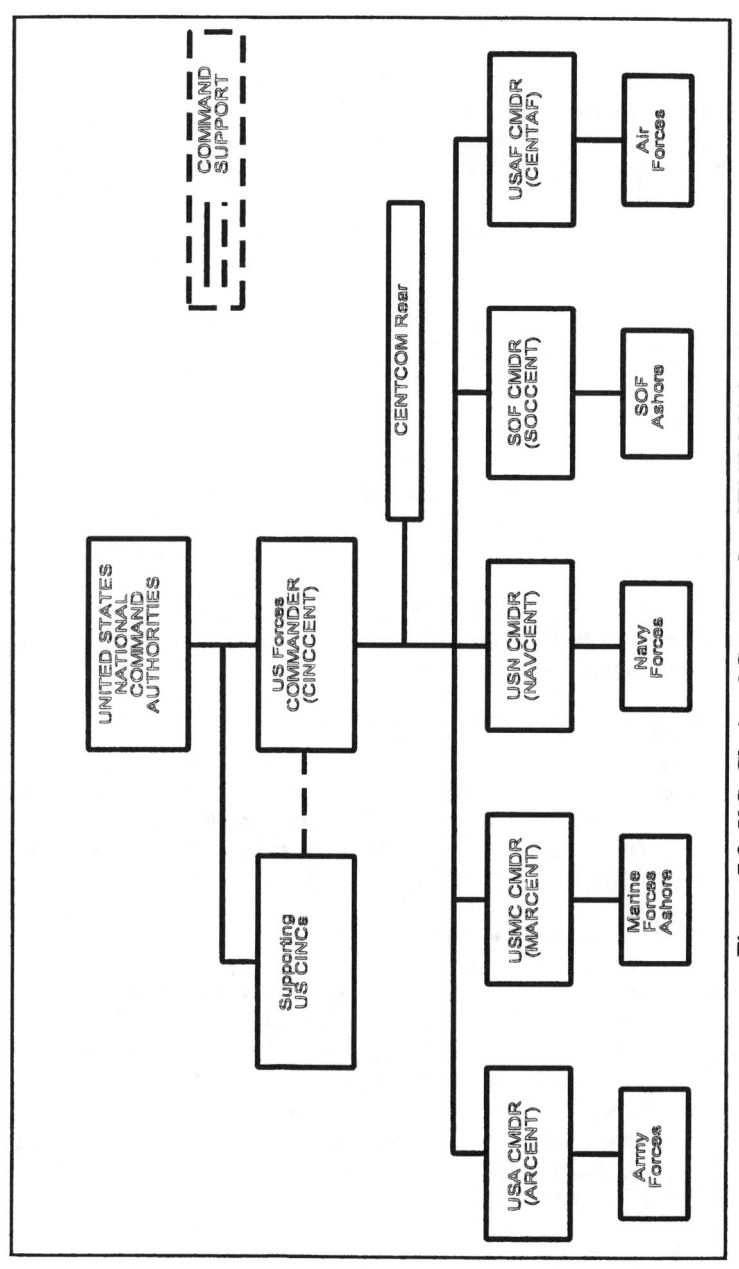

Figure 7.2 U.S. Chain of Command -- CENTCOM AOR

- **Mass**   Mean 9.2

The attack on Hussein's forces in Kuwait was delayed so that sufficient men and material could be deployed to the AOR to ensure the Coalition would be capable of employing overwhelming force at the right time and place. This fits with the American penchant for using material might to minimize casualties. A statement from an officer involved in the gulf operations effectively summarized the use of the principle of mass during the war: "Lots of mass, everywhere in the theater!"

The use of mass in the gulf can be viewed from a number of perspectives, including:
(1) The relative strength in numbers of opposing forces at the beginning of Desert Storm, as illustrated in Tables 5.1 and 5.2 on pages 109 and 110.
(2) The relative strength of those same forces after the air campaign degraded Iraqi units and destroyed their logistics capabilities.
(3) The use of technologically advanced weapons having the stealth and accuracy necessary to compare favorably with the advantages of mass, yet employing significantly fewer numbers.
(4) The employment of AirLand Battle doctrine with the use of a multiplicity of weapon systems from all branches of the services to defeat the Iraqi defensive strategy.

- **Economy of Force**   Mean 8.2

Attempting to properly use the principle of mass while at the same time economizing force is an exacting goal, and one that the U.S. military tends to misapply. It always seems better to go to war with too much power and overwhelm the enemy than to use too little and come up short while sustaining a high casualty rate.

Results of the survey showed this to be the lowest-rated principle of the group. Nevertheless, the United States did apply economy of force during the prosecution of the war. The use of Coalition forces was not simply a political move; it brought in additional combat power from other countries. The use of Arab forces on the Kuwait-Saudi border and the threat of a Marine amphibious landing from the gulf held the attention of a significant number of Iraqi units while the main Coalition attack was launched on the western flank. Directing the major effort against the Republican Guard—the Iraqi center of gravity —minimized less effective operations.

- **Maneuver**   Mean 8.8

Maneuver at the *strategic level* was the first successful application of the principle in response to the Iraqi invasion. The decision and the capability to move large numbers of men and equipment halfway around the world in such a short period of time surely caught Hussein by surprise. His intentions after Kuwait were not known, but his window of opportunity was closed quickly by the rapid political and military response on the part of U.S. leadership and the capabilities of U.S. strategic airlift and sea lift.

At the *operational level*, General Schwarzkopf's "Hail Mary" response of attacking into Iraq and Kuwait on the western flank was the application of maneuver warfare at its finest. "This was absolutely an extraordinary move," the General said in his briefing to the press near the end of the war. "I must tell you, I can't recall any time in the annals of military history when this number of forces have moved over this distance to put themselves in a position to be able to attack."[333] To accomplish this monumental task also required the superb application of the principles of security and surprise to gain an initial advantage over the Iraqi forces.

- **Security**   Mean 8.4

Security was efficiently maintained mainly through the use of technology, satellites, reconnaissance aircraft, special operations, etc. The one failure in security was the surprise Iraqi attack at Al Khafji, which ultimately resulted in an Iraqi defeat.

The highest potential for security failure emanated from the media, a problem that has haunted U.S. military operations for years. The issue essentially falls on a continuum. At one extreme is the right of the public in a democratic society to be informed of circumstances surrounding a crisis during which they may be called upon to sacrifice blood and resources. At the other end of the spectrum is the need to withhold information to protect those involved in military operations. A total suppression of news, however, may work against the military in that public confidence could be lost if adequate and accurate knowledge is not made available. Unfortunately, a reasonable point between the two extremes has not been, and probably never will be, found to the

---

[333] H. Norman Schwarzkopf, "Central Command Briefing," Briefing to the press in Riyadh, Saudi Arabia, on 27 February 1991, cited in *Military Review*, LXXI, no. 9 (September 1991): 97.

satisfaction of all concerned. Survey participants rightly pointed out that CNN and other broadcast networks with their extensive resources and technological capabilities made this principle of war difficult to achieve in the gulf, and will only make it more problematic in future crises.

- **Surprise**   Mean 8.2

In the Gulf War, surprise reigned, beginning with President Bush's strategically unexpected reaction to the invasion of Kuwait. (Who, including Hussein himself, would have been able to predict beforehand the forceful stand the President would take toward Iraqi aggression?) Likewise, the coalition of countries that was formed—and preserved—throughout the conflict, threw not only Hussein, but the world, for a bit of a loop. In the same league as these two milestones was the President's decision to initiate Desert Storm in the absence of further aggressive acts by Hussein, along with his willingness to allow the military to perform without being overly encumbered by excessive political and geographical restraints.

At the operational level, air power decapitated Saddam's intelligence capability and allowed General Schwarzkopf to successfully implement his plan for the ground war—where the deception plan for the campaign worked superbly. In fact, the strategy allowed the main Coalition force to relocate to the western flank and then to drive into the Iraqi rear without being observed. The Republican Guard was totally surprised when faced with the main body of an invading force. At a 27 February 1991 briefing, Schwarzkopf stated:

> We knew that he [Saddam Hussein] had very, very limited reconnaissance means. Therefore, when we took out his air force, for all intents and purposes, we took out his ability to see what we were doing down here in Saudi Arabia. Once we had taken out his eyes, we did what could best be described as the "Hail Mary play" in football.[334]

In essence, by conducting operations to enhance security the Coalition strengthened the surprise element of the ground campaign.

- **Simplicity**   Mean 8.0

---

[334] Schwarzkopf, "Central Command Briefing," 97.

At a news conference on 23 January 1991, General Powell described the U.S. military concept for driving Iraqi forces out of Kuwait: "Our strategy to go after this Army is very, very simple. First we're going to cut it off, and then we're going to kill it."[335]

Although no operational concept for a campaign can be as simple as General Powell made it out to be, commanders following the basic principles of war in the Gulf promoted an unwavering directness toward achieving the stated objectives at the strategic and operational levels. All theater commanders knew the goals, and were therefore able to prepare guidelines and campaign strategies to achieve them in as practical and efficient a manner as possible.

Simplicity at the tactical level had to be approached somewhat differently since the sheer numbers of participants in a limited geographic area—particularly in the air war—made operations extremely complex. Aircraft ingressing and egressing Kuwait and Iraq, timing for air refueling, coordinating search and rescue, changing missions based on real-time battlefield support requirements, and conducting reconnaissance sorties—all around the clock—dictated elaborate planning and attention to detail. The ground campaign had its own set of requirements that necessitated the same skill in planning. A questionnaire comment provides an excellent illustration of how the principle of simplicity was instituted within the context of such elaborate designs: "Planned, prepared, trained, and rehearsed, rehearsed, rehearsed. *Every* soldier knew the plan in detail." Nevertheless, simplicity was rated relatively low in the survey, which was to be expected based on the face value of Gulf War offensive operations, particularly in the air campaign.

Center of Gravity

The Iraqi centers of gravity were specifically identified by the United States in its operational concept and were examined in Chapter 5. Many survey participants and other professionals would have added Saddam Hussein to the list, but that gets back to the argument of how to "take him out" without violating American values or potentially sustaining large numbers of casualties. Some centers of gravity may not be worth the cost of attempting to eliminate them.

---

[335] Colin L. Powell, news conference in Washington, DC, on 23 January, 1991, quoted in Harry G. Summers, *A Critical Analysis of the Gulf War* (New York: Dell Publishing, 1992), 180.

The Angolan War Case

Generally speaking, individuals involved in the Angolan War believed that the principles of war were improperly applied during the prosecution of the war. A number of survey respondents and other individuals interviewed by the author took an extremely *personal* view of the way operations were conducted during the war, not unlike the way many U.S. servicemen viewed the Vietnam War, with a feeling of cynicism and an enduring sense of betrayal.

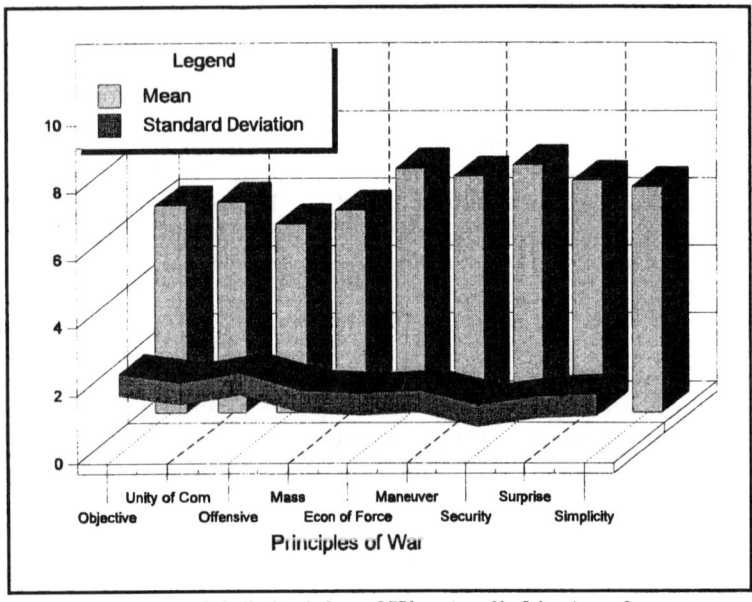

Figure 7.3 Principles of War Applied in Angola

Figure 7.3 provides a graphic portrayal of the statistical mean and standard deviation for each principle as it was perceived to be applied in the Angolan War. The *relatively* low means compare favorably with comments recorded on the questionnaires and contributed in interviews. The means average significantly lower than those found in the Gulf War survey. The large standard deviation indicates a wide disparity in the respondents' views, a feature that will be explored later.

- **Objective**   Mean 6.1

The objective of the war was rated fairly low because no real aim was explicitly stated, and the presumed aim apparently shifted as circumstances dictated. The original objective was to keep UNITA and its base at Mavinga intact by stopping the FAPLA offensive drive from Cuito Cuanavale. Initial South African military successes led to the adoption of further, more ambitious strategic objectives:
  - Curtail ANC and SWAPO use of bases in southern Angola.
  - Eliminate Cuban forces from Africa.
  - Gain a satisfactory solution to the Namibian issue.
  - Remove a military threat from South Africa's northern border.

While these goals were established and understood in the higher reaches of government circles and military headquarters, they were never transmitted to headquarters components outside Pretoria. At the same time, political expediency toward potential negotiations necessitated a restraint be placed on operations in the field as troops were nearing a limited military victory at Cuito Cuanavale.

- **Unity of Command**   Mean 6.2

The chain of command for this campaign was unified but was not considered suitable due to the number of layers (synonymous with the U.S. chain of command supporting the Marines in Beirut) and the propensity for commanders at higher headquarters to bypass intermediate headquarters. Figure 7.4 portrays the chain of command.

The government consisted of the State President, ministers in his Cabinet, the State Security Council, and the top command structure of the SADF. The General Officer Commanding (GOC) of South West Africa (SWA) was charged with directing all military operations

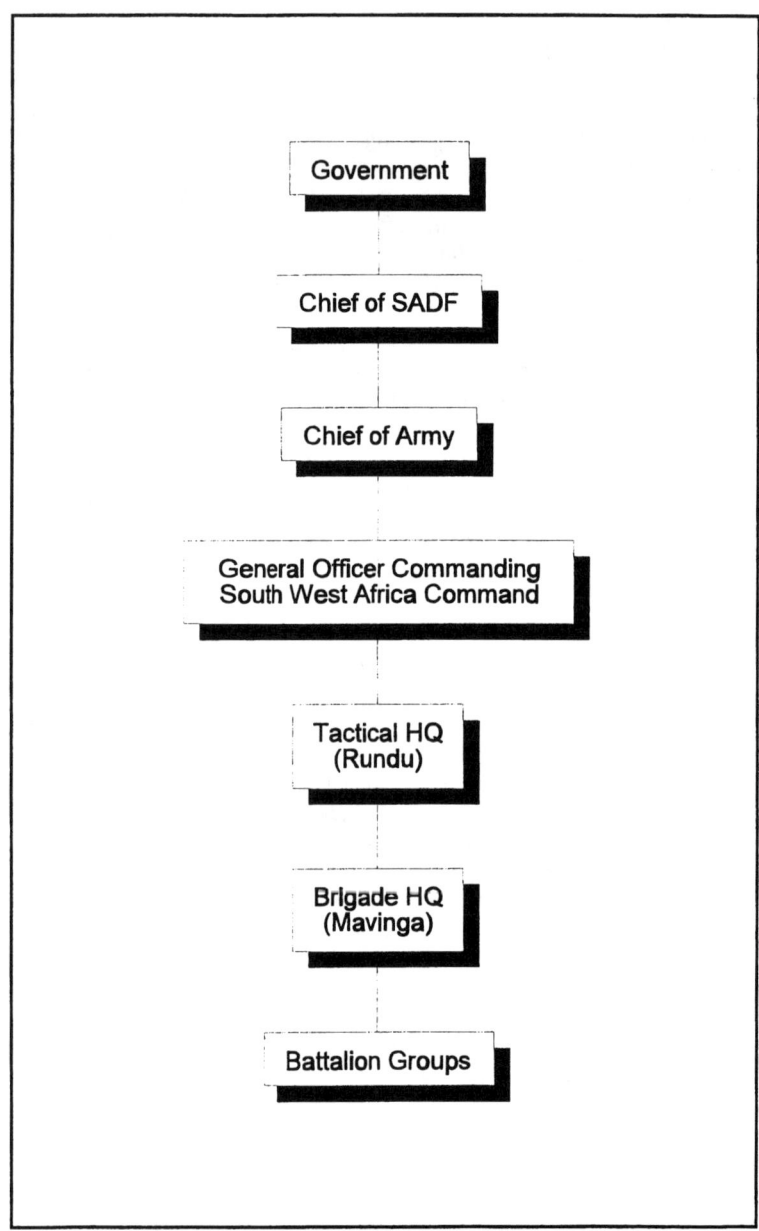

Figure 7.4 SADF Chain of Command

against SWAPO and FAPLA in Namibia and southern Angola. The division commander was at Tactical HQ, located in Rundu, Namibia. The chain of command would have been better suited if the GOC SWA and Tactical HQ had been removed from the command structure since the two had few functional responsibilities and brigade presentations were usually made directly to the Chief of the Army and his staff.

The primary problem relating to the chain of command was higher headquarters involvement at lower levels of command. Several officers commented that "senior political and military officials involved themselves with tactical and even technical matters." The major effect was to limit the freedom of movement and command responsibility of the operational commander at the front as well as to sacrifice time-sensitive tactical initiatives. "The style of command saw C[hief] [of the] Army in person present during the action in which I took part. This resulted in C[hief] [of the] Army taking the responsibility for decisions at unit level."

- **Offensive**   Mean 5.6

The offensive was effectively employed at the strategic, operational, and tactical levels of war for most of the campaign. The FAPLA units were kept off balance, rarely held the initiative, and were defeated on the battlefield. Tactical success in battle led to strategic success at the negotiating table.

The offensive was not, however, employed as effectively as it could have been because of the unwieldy command structure. This arrangement often caused delays in approving timely operations conceived to follow up tactical military successes; hence the initiative was lost. Additionally, a number of serious shortcomings hindered offensive operations. These included:
- Piecemeal commitment of additional forces as involvement in Angola expanded; the force was too small to efficiently accomplish assigned tasks and allow for appropriate reserves.
- Restrictions placed on offensive operations due to political considerations.
- An extreme concern to limit casualties and loss of equipment.
- The requirement to rotate national servicemen after each operation because time toward their duty obligation had expired.
- Change of commanders after each operation.
- Long lines of communication resulting in logistical support problems, especially with respect to fuel, tank tracks, G5 barrels, and ammunition.

- Lack of friendly air cover that allowed the enemy to gain air superiority.

These deficiencies led to the final operation being viewed as a failure by the men charged with driving FAPLA troops from the east bank of the Cuito River. The war, in effect, became personal, and although it led to strategic successes, it was seen in the eyes of the men fighting at the tactical level as a cause lost by the staff at higher headquarters.

- **Mass**   Mean 6.0

The application of mass in the Angolan campaign was done at the tactical level and was properly employed within the limits of the political constraints placed on the operations: "In all cases, the bulk of the combat power available to the tactical commander (including air power) was employed at the decisive point at the decisive time." FAPLA troops were not in a position to mutually support each other so it was possible to concentrate SADF units against weak points in the defense. Small units of SADF and UNITA forces also conducted operations west of Cuito Cuanavale, pinning down large FAPLA troop concentrations that otherwise would have been used in the main battle area east of Cuito.

The size of the SADF units in Angola was limited for political reasons, which hindered a better application of the principle of mass. The belief in Pretoria was that too strong an offensive thrust by a large force would cause the Cubans and Russians to panic and would lead to the deployment of more Cuban troops to Angola. As a result, the bulk of SADF troops was never employed during the war.

Toward the end of the war FAPLA troops were allowed to mass in and around Cuito Cuanavale. Terrain, a small force, and lack of heavy equipment such as tanks would have prevented the SADF from pursuing the offensive if the decision had been made to do so.

- **Economy of Force**   Mean 7.2

South Africa never used its full military capability in Angola. From an economy of force viewpoint, this judicious use of limited combat power from a strategic and operational perspective was sound strategy based on the political situation. The FAPLA offensive against Mavinga was stopped with few resources expended and minimal casualties. Many operations were conducted by small SADF and UNITA units to ambush convoys and troop columns. Infiltration operations were carried out to direct artillery strikes on high-value, high-priority targets. The G5 and G6 artillery alone maximized the concept of using minimum essential combat power against an objective. On several occasions this artillery was used almost exclusively to destroy FAPLA troop concentrations and break up the cohesion of their combat assaults.

Minimum force in theater did, however, have a negative impact. Force levels were actually too small to correctly apply this principle because all forces had to be employed in the main action. No forces were therefore available to dissipate enemy strength. This condition was pivotal in limiting the success of Operation Packer, the final operation of the war.

- **Maneuver**   Mean 7.0

At the operational level the principle of maneuver was never fully utilized in the Angolan War, with good reason. If large SADF units attacked in other areas of southern Angola (for instance, west of Cuito Cuanavale, as several tactical commanders would have liked), then the conflict could easily have escalated. The MPLA would have considered the move as a threat to its major logistics base at Menongue and would have certainly called for more Cuban and Russian support. The question here is: Where does one draw the line in *restricting* logical military maneuvers based on the potential for a *perceived* response?

Conversely, the concept of operations made very good use of maneuver at the tactical level. The South Africans were supplied with vehicles especially designed to operate in the "bush," and their units were particularly well structured for mobile operations over long distances. The troops were trained in maneuver warfare even though the terrain and vegetation of southern Africa made this type of operation difficult. When accomplished with small units, the operations proved highly effective, resulting in operational successes and enemy losses far in excess of what would be considered normal given the enemy's numerical superiority.

- **Security**   Mean 7.3

Security was effectively employed by the SADF at the tactical level during most of the campaign, resulting in considerable operational success. It was accomplished mainly through the combined efforts of UNITA light forces and South African liaison and special forces teams. UNITA provided a security screen and guaranteed a safe area for South African forces at all times. SADF infiltration operations maintained constant surveillance in FAPLA territory. Electronic warfare was effective in intercepting FAPLA transmissions, while the SADF's own equipment provided secure voice capabilities.

As SADF and UNITA forces approached Cuito Cuanavale for the final offensive push of the war, security apparently began to break down, however. "As clashes took place nearer to FAPLA's home ground at Cuito Cuanavale, their EW [electronic warfare] ability increased to such an extent that they probably knew of every impending attack soon after the SA deployment started." This weakness certainly contributed to the operational ineffectiveness during the final assaults near the Cuito River.

- **Surprise**   Mean 6.8

At the strategic level, the FAPLA/Cuban force was apparently surprised that the SADF and UNITA were able to mobilize quickly enough to counter the advance on Mavinga and was unprepared for "the ferocity of the counter-attack." The SADF's escalation of the war by deploying heavy mechanized forces such as tanks was also not expected.

The use of surprise throughout the campaign was successfully employed by the SADF at the tactical level until the final battles at the Cuito River. Night movements, rapid deployments through rough terrain, speed of pursuit, communications security, use of long-range artillery, and deception measures all contributed to successful ambushes, entrapments, and unanticipated offensive operations. It was only near Cuito, where options for direction of attack, speed, and deception were reduced, that the principle of surprise was not used effectively.

- **Simplicity**   Mean 6.7

"Tactical plans were always simple, clearly understood and easily executed. This was the result of standard doctrine and training and was not unique to the operation." The unfavorable terrain also had a considerable influence on planning, requiring designs to be relatively uncomplicated. The major weakness of this principle came from the coordination procedure. There were too many levels in the decision-making, planning, and controlling process.

Center of Gravity

There was no consensus as to what the center of gravity was on the FAPLA side during the Angolan war. Respondents identified the following as likely candidates: specific FAPLA units; the town of Cuito Cuanavale; the airport at Cuito Cuanavale; Luanda, the capital of Angola; the enemy's will to fight and its commanders' ability to make the right decisions under pressure; the road between Cuito and the logistics base at Menongue; and finally, the town of Menongue itself.

The most insightful response identified the town of Menongue as the center of gravity because it was the main logistics base, contained a railhead and the region's major airport, and was FAPLA's principal command center. Although there initially were plans to attack Menongue and use it as a point to launch an attack on Cuito from the west, the option was ruled out for the following reasons:
- The South African Cabinet would not allow SADF troops to enter and hold any Angolan town, primarily for foreign policy reasons, but also because of concerns over public reaction in South Africa.
- The Chief of the SADF, General Geldenhuys, would not approve any plan that would lead to direct contact with Cuban troops. He believed that a Cuban defeat would lead to more Cubans being sent to Angola rather than being withdrawn.

Political and strategic reasons therefore prevented the proper exploitation of this center of gravity.

Cause and Effect

It became evident during interviews with military officers who fought in Angola that many took the war personally and blamed the establishment for what they considered to be a less than satisfactory conclusion to military operations. This being the case, was there a potential problem concerning their objectivity in responding to the principles of war survey? What contributed to the relatively low scores on each principle? Did the respondents rate the principles low because they were actually applied inappropriately? Or, did they rate them low because the SADF failed to take the final objective (the east bank of the Cuito River) at the end of operations in Angola? Which was the cause and which was the effect?

In order to assess this concern, respondents were divided into two groups: the fighters—military officers who fought in Angola and who accounted for 60 percent of the sample population of 20; and the non-fighters—individuals who were on higher headquarters staffs and in other germane professions at the time of the war and constituted 40 percent of the sample population. The proper application of the principles of war was analyzed for each group. Figure 7.5 displays the results.

As the graph illustrates, the non-fighters rated the appropriate use of most principles somewhat higher than the fighters. This can be explained through one of two reasons:

- The fighters were too personally involved to be totally objective, and gave lower marks because of their perceived failure in Angola.
- Those at higher headquarters and elsewhere had a better understanding of the strategic implications of the political constraints placed on military operations and in the end believed the military operations contributed directly to successful negotiations.

Proper application of the principles of war was rated lower across the board for the Angolan War than the Gulf War. And the ratings would remain lower even if the fighters were factored out of the analysis. This being the case, it is safe to say that personal experience during the Angolan War did not significantly influence the final outcomes of this research.

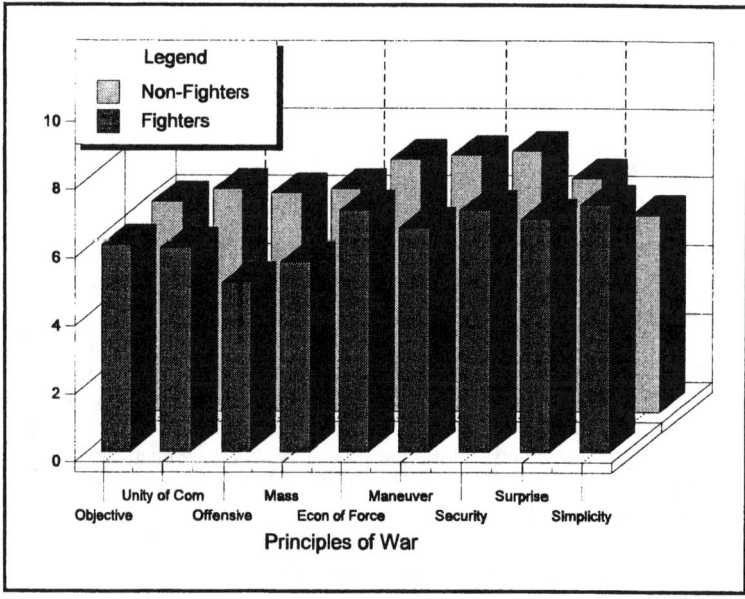

Figure 7.5 Principles of War Analysis—
Fighters vs. Non-fighters

Comparing the Two Cases

The principles of war were perceived to be more properly applied in the Persian Gulf than in Angola across the principles spectrum. Only "economy of force" was judged to be applied effectively in a relatively equal manner in both wars. (See Figure 7.6.)

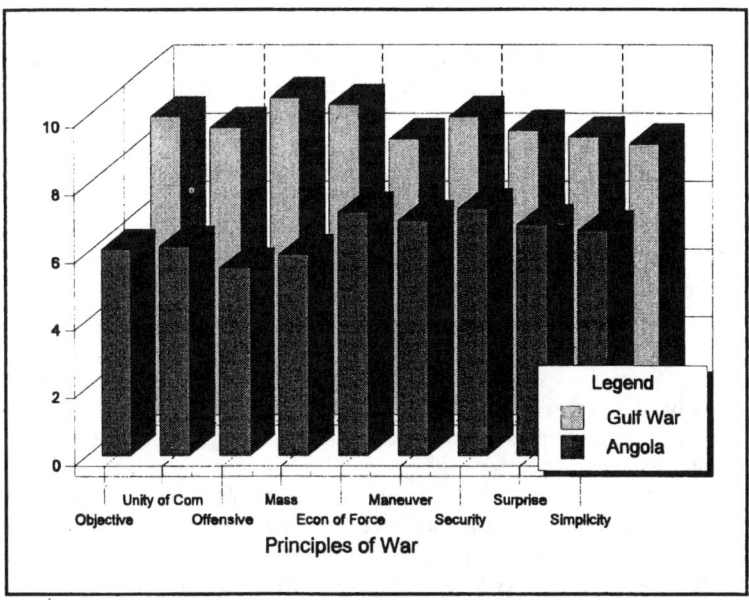

Figure 7.6 Comparing the Gulf War with Angola

The American responses extolled the competence of the U.S. leadership, efficiency of the campaign planning, and excellence of operational execution in the Gulf War. The South African responses, on the other hand, were extremely critical of the way the Angolan War was conducted and the way the principles of war were employed.

A slight surprise came in the comparison of the two standard deviations. It is surprising then, that the standard deviations for the principles of economy of force, security, surprise, and simplicity were relatively equal or smaller for Angola, indicating closer agreement on the part of the respondents on those principles in the Angolan survey than in the gulf survey. (See figure 7.7.) Considering the survey population for the Angolan War (i.e., fighters and non-fighters), one would have expected a relatively large variance across the board.

The "Principles of War" Questionnaire was not designed as an in-depth statistical analysis of the Gulf and Angolan Wars. Its purpose was to establish a comparative trend in the use of the principles of war and results stemming from their application.

In fact, the survey results indicate a correlation and help to substantiate that:

- The use of the principles of war in the prosecution of war remains a valid concept, even under contemporary, high-technology, high-intensity combat.
- Relative success or failure of warfare can be related to the application of the principles of war in the conduct of a war.

In the gulf, for example, the *military* victory was one of the most one-sided triumphs in recorded history due primarily to the application of technologically superior weapons and using the principles of war as a basis for targeting and maneuver. In addition, the campaign plan itself incorporated the principles of war and was considered almost flawless, based on the impressive results each phase of the plan achieved. Finally, the survey results reflected the operational results of the war—each of the principles of war was rated as being properly employed during the campaign.

Similarly, military results in the Angolan War and the associated lower means and comments obtained from the principles survey support the above hypotheses, with some limitations. Namely, the use of high-technology G-5 and G-6 long-range howitzers enabled the SADF to properly apply several principles of war, such as economy of force and surprise. However, survey results revealed questionable applications of most principles when "decision making" was the operative factor in the conduct of the war.

Implementation of the principles of war in decision making requires judicious handling. The principles are meant to be used as a bridge to help alleviate the cultural differences of the civilian and military decision makers so that they can address a crisis from the same perspective. Each leader must understand what would be required to: support military objectives, identify necessary resources, and assess operational considerations. This allows each civilian leader and military commander throughout the chain of command to know what is expected when conducting a campaign should the situation induce a military response. The principles of war *should not be used* as an inflexible checklist! The literature is very clear on this, and the position is absolutely on target. Each crisis situation is different and would have to be acted upon based on the particular circumstances at that time. The principles, used as a guide, help narrow the focus on what should be considered if a military response is required.

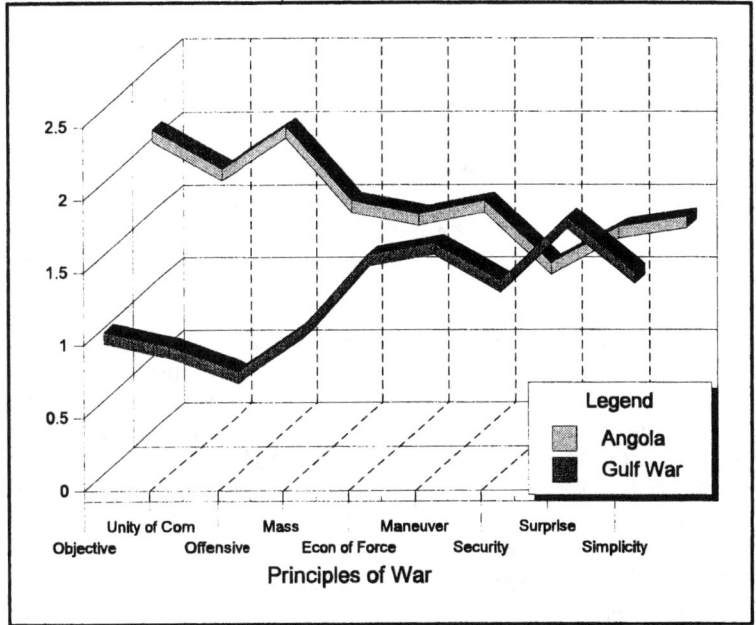

Figure 7.7 Standard Deviations Compared

Finally, John Collins places the use of the principles of war in the proper perspective:

[S]uccessful strategists never knowingly violate the Principles of War unless they first evaluate the risks and estimate expenses. Readers who apply this yardstick to any conflict or period of international tension in history must conclude that, critics notwithstanding, the Principles of War are utilitarian and they do make sense. The record shows that winners, by and large, took heed of the Principles. The losers, discounting those who were overcome by sheer weight of manpower and material, by and large did not.[336]

---

[336] Collins, 28.

A Special Case: War Termination

History has shown, and the principles of war survey exhibits, that success on the battlefield does not necessarily mean that the desired political outcome will be achieved. From a military perspective, results of the war in Angola were not gratifying, as the survey illustrated. The war actually resulted in a military stalemate. However, war termination as a direct result of those military operations produced a dramatic advance in the political fortunes of southern Africa.

On the other hand, the Persian Gulf War, with all its unequivocal battlefield successes, is a perfect illustration of the paradoxical consequences of war termination. Saddam Hussein's military forces were crushed by the Coalition forces, yet Hussein remained in power. Residual forces from his defeated elite Republican Guard remained potent enough to turn away attempts by the Kurds and Shiite Moslems to oust him from power. A myriad of atrocities were committed upon the rebel forces and civilians alike during that struggle for power. Kurdish refugees died by the thousands from starvation and unabated winter weather as they attempted to escape to Turkey and Iran after the war. On the international scene, a U.S. effort to organize a peace conference between Israel and its Arab neighbors was rebuffed. The liberated Kuwaiti people were no closer to a democratic form of government than they were before the war. In short, the only step toward a stable peace in the region emanating from the war was the near destruction of the fourth largest army in the world and its associated means of mass destruction.[337]

The lesson to be learned is that military conquest may at times be the easiest part (once the decision is made to fight) of conducting policy in the pursuit of national interests. The difficulty lies in managing the political morass after conflict termination. Results achieved by employing the political instrument of national power will, in the end, usually determine the success or failure of national security policy at the end of the day.

---

[337] David V. Nowlin and Ronald J. Stupak, "War Termination and the Persian Gulf War: Getting-In, Getting-Out, and Getting It Right," *American Society for Public Administration: SNSDA* 11, no.1 (Summer 1991): 7.

Appendix: Survey Questionnaire

"PRINCIPLES OF WAR" QUESTIONNAIRE

Your position during the 1987/88 War in Angola was critical to the success or failure of that conflict and the future relations of South Africa with its immediate neighbors and the rest of the world. The purpose of this questionnaire is to obtain your perspective on how the "principles of war" were employed in the conduct of the war. Please base your views in relation to the following hierarchy of warfare:

Strategic Level: The level of war at which a nation determines national or alliance security objectives and develops and uses national resources to accomplish these objectives (State President, Cabinet, Parliament, SADF Headquarters, etc.).

Operational Level: The level of war at which campaigns and major operations are planned, conducted, and sustained to accomplish strategic objectives within theaters or areas of operations (Intermediate Headquarters).

Tactical Level: The level of war at which battles and engagements are planned and executed to accomplish military objectives assigned to tactical units or task forces.

Place an "X" in the appropriate box for each principle of war that more closely describes its employment effectiveness during the 1987/88 Angolan War. Attempt to answer the questions at the **strategic** and/or **operational** levels of war rather than getting bogged down in detailed operations at the tactical level.

Objective: Direct every operation toward the achievement of an objective that is clearly defined, attainable, and decisive.

1	2	3	4	5	6	7	8	9	10

UNCLEAR   CLEARLY DEFINED

What did you consider to be the strategic objective(s) during the war? The theater objectives? What did you hope to achieve at war termination? Would you consider the area/world a better place at war termination?

Unity of Command: For every objective, ensure unity of effort under one responsible commander. Other components of unity of effort are common objectives, coordinated planning, and trust.

1	2	3	4	5	6	7	8	9	10

NOT CLEAR   WELL DEFINED

Describe the chain of command. What did you consider its strengths and weaknesses?

How effective was the chain of command? What did you consider its strengths and weaknesses?

Offensive: Seize and exploit the initiative to set the terms of the engagement. Military victory requires decisive action, often in coordination with defensive action. The aim is to generate an operational momentum to which enemies cannot successfully react, depriving them of freedom of action.

Appendix: Survey Questionnaire

1	2	3	4	5	6	7	8	9	10

IMPROPERLY EMPLOYED   PROPERLY EMPLOYED

How effectively was the offensive employed in support of the objectives?

What actions facilitated offensive operations? What actions hindered these operations?

Mass: Concentrate sufficient combat power at the correct time and place to achieve decisive results. At the same time, force enemies to dissipate their strength so that they cannot concentrate.

1	2	3	4	5	6	7	8	9	10

IMPROPERLY EMPLOYED   PROPERLY EMPLOYED

Describe how the principle of mass was properly/improperly employed.

Economy of Force: Allocate minimum essential combat power to secondary efforts in order to dissipate enemy strength and to achieve superiority in the area where decision is sought.

1	2	3	4	5	6	7	8	9	10

IMPROPERLY EMPLOYED   PROPERLY EMPLOYED

How was economy of force used to further the war efforts? How was it improperly employed or not used?

Maneuver: Place the enemy in a position of disadvantage through the flexible use of combat power.

1	2	3	4	5	6	7	8	9	10

IMPROPERLY EMPLOYED   PROPERLY EMPLOYED

What actions supported the use of maneuver during the war and what actions hindered its application? How was it effectively/ineffectively employed?

Security: Prevent the enemy from achieving an unexpected advantage. Take continuous, positive action to prevent surprise, to retain flexibility, and to preserve freedom of action.

1	2	3	4	5	6	7	8	9	10

IMPROPERLY EMPLOYED   PROPERLY EMPLOYED

How was security employed to further the war effort? Where did it fail?

Surprise: Take action against enemies at times, places, and in manners for which they neither are prepared nor expect.

1	2	3	4	5	6	7	8	9	10

IMPROPERLY EMPLOYED   PROPERLY EMPLOYED

How did the principle of surprise aid in/detract from successfully attaining the war's objectives?

Appendix: Survey Questionnaire

Simplicity: Issue clear, concise, uncomplicated plans, orders, and/or guidance.

1	2	3	4	5	6	7	8	9	10

IMPROPERLY EMPLOYED   PROPERLY EMPLOYED

How was this accomplished or not?

The enemy's **center of gravity** is defined as "the hub of all power and movement on which everything depends . . . . The point against which all energies should be directed."

What did you consider to be the enemy's center of gravity? Was it properly exploited?

Glossary

(SOURCE: U.S. Department of Defense, Joint Staff, *Joint Pub 1-03; Department of Defense Dictionary of Military and Associated Terms* (Washington, DC: Department of Defense, The Joint Chiefs of Staff, 1 December 1989).

Chain of command: The succession of commanding officers from a superior to a subordinate through which command is exercised.

Crisis: An unstable or crucial time or state of affairs in which a decisive change is pending, especially one with the distinct possibility of a highly undesirable outcome.

Estimate of the situation: A logical process of reasoning by which a commander considers all the circumstances affecting the military situation and arrives at a decision as to the course of action to be taken in order to accomplish his/her mission.

National objectives: Those fundamental aims, goals, or purposes of a nation, as opposed to the means of seeking these ends, toward which a policy is directed and efforts and resources of the nation are applied.

National security policy: A broad course of action or statements of guidance adopted by the government at the national level in pursuit of national objectives.

National security strategy: A realistic approach to dealing with threats to a nation's security that reflects national interests and presents a broad plan for achieving national objectives.

<u>National strategy</u>: The art and science of developing and using political, economic, and psychological powers of a nation, together with its armed forces, during peace and war, to secure national objectives.

<u>Operational level of war</u>: The level of war at which campaigns and major operations are planned, conducted, and sustained to accomplish strategic objectives within theaters or areas of operations.

<u>Strategy</u>: The art and science of developing and using political, economic, psychological, and military forces as necessary during peace and war, to afford the maximum support of policies, in order to increase the probabilities and favorable consequences of victory and to lessen the chances of defeat.

<u>Strategic level of war</u>: The level of war at which a nation or group of nations determines national or alliance security objectives and develops and uses national resources to accomplish those objectives.

<u>Tactical level of war</u>: The level of war at which battles and engagements are planned and executed to accomplish military objectives assigned to tactical units or task forces.

Selected Bibliography

Act to Establish an Executive Department, to Be Denominated the Department of War, August 7, 1789. 1, Stat. 7, sec. 1 (1789). Quoted in Frederick C. Mosher, ed. Basic Documents of American Public Administration, 1776-1950. New York: Holmes & Meier Publishers, Inc., 1976.

Allison, Graham T. Essence of Decision. Boston: Little, Brown and Company, 1971.

Alonso, Rod. "The Air War." In Military Lessons of the Gulf War, ed. Bruce W. Watson, 61-80. 2nd rev. ed. London: Greenhill Books, 1993.

Antal, John F., Major, U.S. Army. "Maneuver Versus Attrition, a Historical Perspective." Military Review 10 (October 1992): 21-33.

Art, Robert J. and Kenneth N. Waltz, eds. The Use of Force, International Politics and Foreign Policy. 2nd ed. New York: University Press of America, 1983.

Atkeson, E.B., Maj. Gen., U.S. Army (Ret.). The Final Argument of Kings, Reflections on the Art of War. Fairfax, VA: Hero Books, 1988.

Babbage, Charles. On the Economy of Machinery and Manufactures. London: Charles Knight, Pall Mall, East, 1835; reprint, New York: Augustus M. Kelly, Bookseller, 1963.

Barnard, Chester I. The Functions of the Executive. Cambridge, MA: Harvard University Press, 1938.

Barnet, Richard J. Roots of War: The Men and Institutions Behind U.S. Foreign Policy. Baltimore: Penguin Press, 1972, 48-49, Quoted in Charles W. Kegley, Jr. and Eugene R. Wittkopf. American Foreign Policy, Pattern and Process. 3rd ed. New York: St. Martin's Press, Inc., 1987.

Beaufre, André. An Introduction to Strategy. London: Faber and Faber, 1965.

Berridge, Geoffrey R. "Diplomacy and the Angola/Namibia Accords." International Affairs 65, no. 3 (Summer 1989): 463-479.

Betts, Richard K. Soldiers, Statesmen, and Cold War Crises. Cambridge, MA: Harvard University Press, 1977.

Blake, Thomas G. The Shield and the Storm. Jostens Inc., 1992.

Blechman, Barry M. and Stephen S. Kaplan. Force without War. Washington, DC: The Brookings Institution, 1978.

Bonaparte, Napoleon. Memoirs, Vol. II. London: H. Colburn and Co., 1823-24. Quoted in John M. Collins. Grand Strategy, Principles and Practices. Annapolis, MD: Naval Institute Press, 1973.

Bond, Brian and Martin Alexander. "Liddell Hart and De Gaulle: The Doctrines of Liability and Mobile Defense." In Makers of Modern Strategy From Machiavelli to the Nuclear Age, ed. Peter Paret, 598-623. Princeton: Princeton University Press, 1986.

Braestrup, Peter, ed. Vietnam as a History. Washington, DC: University Press of America, Inc., 1984.

Bridgland, Fred. The War For Africa: Twelve Months That Transformed a Continent. Gibraltar: Ashanti Publishing Limited, 1990.

Selected Bibliography

_____. "Angola and the West." In <u>Challenge: Southern Africa Within the African Revolutionary Context</u>, ed. Al J. Venter, 117-145. Gibraltar: Ashanti Publishing Limited, 1989.

_____. <u>Jonas Savimbi: A Key to Africa</u>. Johannesburg: Macmillan South Africa (Publishers) (Pty.) Ltd., 1986.

Brodie, Bernard. <u>Strategy and National Interests, Reflections for the Future</u>. New York: National Strategy Information Center, Inc., 1971.

_____. <u>Strategy in the Missile Age</u>. Princeton: Princeton University Press, 1965.

_____. <u>War & Politics</u>. New York: Macmillan Publishing Co., Inc., 1973.

Brodie, Bernard and Fawn M. Brodie. <u>From Crossbow to H-Bomb</u>. Bloomington, IN: Indiana University Press, 1973.

Bronowski, J. and Bruce Mazlish. <u>The Western Intellectual Tradition</u>. New York: Harper & Row, Publishers, 1975.

Bush, George H. "U.S. Military Power Must Help Promote Peace." Speech presented at the Military Academy at West Point on 5 January 1993.

Carpenter, Ted G., ed. <u>Collective Defense or Strategic Independence?</u> Lexington: D.C. Heath and Company, 1989.

Central Intelligence Agency. <u>The World Fact Book 1991</u>. Washington, DC: U.S. Government Printing Office, 1991.

Chandler, Ralph C., and Jack C. Plano. <u>The Public Administration Dictionary</u>. New York: Macmillan Publishing Company, 1986.

Cimbala, Stephen J. and Keith A. Dunn, eds. <u>Conflict Termination and Military Strategy: Coercion, Persuasion, and War</u>. Boulder, CO: Westview Press, 1987.

Clausewitz, Carl Von. On War. Edited and Translated by Michael Howard and Peter Paret. Princeton, NJ: Princeton University Press, 1976.

Collins, John M. Grand Strategy, Principles and Practices. Annapolis, MD: Naval Institute Press, 1973.

The Communist International 4. no. 2 (15 December 1928). Quoted in Morgan Norval. Inside the ANC: The Evolution of a Terrorist Organization. Washington, DC: Selous Foundation Press, 1990.

Craig, Gordon A. "Delbrück: The Military Historian." In Makers of Modern Strategy, From Machiavelli to the Nuclear Age, ed. Peter Paret, 326-353. Princeton, NJ: Princeton University Press, 1986.

Crocker, Chester A. High Noon in Southern Africa: Making Peace in a Rough Neighborhood. New York: W.W. Norton & Company, 1992.

_____. South Africa's Defense Posture: Coping with Vulnerability. The Washington Papers. Vol. 9, no. 84. Beverly Hills, CA: Sage Publications, Inc., 1981.

Davenport, T.R.H. South Africa, A Modern History, 3rd ed. Toronto: University of Toronto Press, 1987.

Davis, Vincent. "The Evolution of Central U.S. Defense Management." In Reorganizing America's Defense, Leadership in War and Peace, eds. Robert J. Art, Vincent Davis, and Samuel P. Huntington, 149-167. Washington, DC: Pergamon-Brassey's International Defense Publishers, 1985.

De Lupis, Ingrid D. The Law of War. Cambridge: Cambridge University Press, 1987.

del Pino Diaz, General Rafael. Interview by Fred Bridgland, December 1987. Washington, DC. Quoted in Fred Bridgland. "Angola and the West," in Challenge: Southern Africa Within the African Revolutionary Context, 132. ed. Al J. Venter. Gibralter: Ashanti Publishing Limited, 1989.

Dimock, Marshal E. "The Criteria and Objectives of Public Administration." in The Frontiers of Public Administration, ed. John M. Gaus, Leonard D. White, and Marshall E. Dimock, 116-133. Chicago: The University of Chicago Press, 1936; reissue, New York: Russell & Russell, 1967.

Douhet, Giulio. The Command of the Air. Translated by Dino Ferrari. New York: Coward-McCann, Inc., 1942; reprint, Washington, DC: Office of Air Force History, 1983.

Drew, Dennis M. and Donald M. Snow. Making Strategy, An Introduction to National Security Processes and Problems. Maxwell Air Force Base, AL: Air University Press, 1988.

Dunn, Keith A. "The Missing Link in Conflict Termination Thought: Strategy." in Conflict Termination and Military Strategy: Coercion, Persuasion, and War, ed. Stephen J. Cimbala and Keith A. Dunn, 175-193. Boulder, CO: Westview Press, 1987.

Eccles, Henry E. Logistics in the National Defense. Westport, CT: Greenwood Press, 1959.

_____. "Strategy: The Theory and Application." in Military Strategy: Theory and Application, ed. Colonel Arthur F. Lykke, Jr. (Ret.), 36-42. Carlisle Barracks, PA: U.S. Army War College, 1989.

Eisenhower, Dwight D. Crusade in Europe. New York: Da Capo Press, 1977.

Fayol, Henri. General and Industrial Management. Translated by Constance Storrs. London: Pitman Publishing, 1949; reprint, London: Pitman Publishing, 1972.

Frankel, P.H. Pretoria's Praetorians: Civil-Military Relations in South Africa. Cambridge: Cambridge University Press, 1984.

Freedman, Lawrence. The Evolution of Nuclear Strategy. New York: St. Martin's Press, 1983.

Freidman, Norman. Desert Victory, The War for Kuwait. Annapolis: Naval Institute Press,1991.

Geldenhuys, Deon. The Diplomacy of Isolation, South African Foreign Policy Making. Johannesburg: Macmillan South Africa (Publishers) (Pty.) Ltd., 1984.

George, Alexander L. Presidential Decisionmaking in Foreign Policy. Boulder, CO: Westview Press, 1980.

Gilbert, Felix. "Machiavelli: The Renaissance of the Art of War." in Makers of Modern Strategy From Machiavelli to the Nuclear Age, ed. Peter Paret, 11-31. Princeton: Princeton University Press, 1986.

Goodpaster, Andrew J. For the Common Defense. Lexington: D.C. Heath and Company, 1977.

Gray, Colin S. The Geopolitics of Super Power. Lexington, KY: The University Press of Kentucky, 1988.

Grest, Jeremy. "The South African Defence Force in Angola." in War and Society: The Militarisation of South Africa, ed. Jacklyn Cock and Laurie Nathan, 116-132. Claremont, S.A.: David Philip, Publisher (Pty.) Ltd., 1989.

Gulick, Luther. "Notes on the Theory of Organization." in Papers on the Science of Administration, ed. Luther Gulick and L. Urwick, 3-45. New York: Institute of Public Administration, 1937.

Halperin, Morton H. National Security Policy-Making. Lexington: D.C. Heath and Company, 1975.

Harmon, Michael M. and Richard T. Mayer. Organization Theory for Public Administration. Glenview, IL: Scott, Foresman and Company, 1986.

Hart, B.H. Liddell. Strategy. 2nd revised ed. New York: NAL Penguin Inc., 1974.

Heitman, Helmoed-Römer. War in Angola, The Final South African Phase. Gibraltar: Ashanti Publishing Limited, 1990.

Hobkirk, Michael D. The Politics of Defence Budgeting. Washington, DC: National Defense University Press, 1983.

Horner, Lt. General Charles A. "The Air Campaign." Military Review LXXI, no. 9 (September 1991): 16-27.

Huntington, Samuel P. American Military Strategy. No. 28, Policy Papers in International Affairs. Berkeley: University of California Press, 1986.

_____. The Soldier and the State. New York: Vintage Books, 1959.

Hurley, Mathew M. "Saddam Hussein and Iraqi Air Power." Airpower Journal 4 (Winter 1992): 4-16.

Hursch, James A., ed. Theories of International Relations. Washington, DC: National Defense University Press, 1990.

Industrial College of the Armed Forces. Military Power and Strategy: Force Determination. Faculty Guide, Academic Year 1989-1990.

Janis, Irving L. Groupthink. 2nd ed. Boston, MA: Houghton Mifflin Company, 1982.

Jomini, Baron Henri. The Art of War. Translated by Capt. G.H. Mendell and Lieut. W.P. Craighill. Philadelphia, PA: Lippincott & Co., 1862; reprint, Westport, CT: Greenwood Press, 1971.

_____. Quoted in J.D. Little, ed. "Jomini and His Summary of the Art of War." Military Strategy: Theory and Application, 94-122, ed. Colonel Arthur F. Lykke, Jr. (Ret.). Carlisle Barracks, PA: U.S. Army War College, 1989.

Kegley, Charles W., Jr. and Eugene R. Wittkopf. American Foreign Policy, Pattern and Process. 3rd ed. New York: St. Martin's Press, 1987.

Kennedy, Paul. The Rise and Fall of the Great Powers. New York: Vintage Books, 1987.

Kiljunen, Kimmo. "The Ideology of National Liberation." In Namibia: The Last Colony, ed. Reginald Green, Marja-Liisa Kiljunen, and Kimmo Kiljunen, 145-171. Burnt Mill, U.K.: Longman Group Limited, 1981.

Kinnard, Douglas. The War Managers. Hanover, NH: University Press of New England, 1977; reprint, New York: Da Capo Press, 1991.

Kissinger, Henry A. Nuclear Weapons and Foreign Policy. New York: Harper & Row, 1957.

Klinghoffer, Arthur Jay. "Soviet Union and Superpower Rivalry." In African Security Issues: Sovereignty, Stability, and Solidarity, ed. Bruce E. Arlinghaus, 1938. Boulder, CO: Westview Press, 1984.

Kronenberg, Philip S., ed. Planning U.S. Security. Washington, DC: National Defense University Press, 1981.

Kruys, George. "Fallacies Are a Matter of Twisted Facts." Paratus 45 (April 1993): 18-22.

Laffin, John. War Annual 2: A Guide to Contemporary Wars and Conflicts. London: Brassey's Defence Publishers, 1987.

LeMay, General Curtis E. Air Force Manual 1-1, Basic Doctrine. Washington, DC: Department of the Air Force, 1984, frontispiece. Quoted in Joint Pub 1, Joint Warfare of the US Armed Forces. Washington, DC: Department of Defense. The Joint Chiefs of Staff, 11 November 1991.

Lord, Carnes and Frank R. Barnett, eds. Political Warfare and Psychological Operations. Washington, DC: National Defense University Press, 1989.

Luttwak, Edward N. Strategy, The Logic of War and Peace. Cambridge, MA: The Belknap Press of Harvard University Press, 1987.

Lykke, Arthur F., ed. Military Strategy: Theory and Application. Carlisle Barracks, PA: United States Army War College, 1989.

Mahan, A.T. The Influence of Sea Power Upon History, 1660-1783. 5th ed. New York: Dover Publications, Inc., 1987.

Malan, Colonel Jan P., Officer Commanding, 4SAI. 1993. Interview by author, 30 August, Upington, South Africa.

March, James G. and Herbert A. Simon. Organizations. New York: John Wiley & Sons, 1958.

Marcum, John A. The Angolan Revolution, Volume II, Exile Politics and Guerrilla Warfare (1962-1976). Cambridge, MA: The MIT Press, 1978.

Matthews, Lloyd J. and Dale E. Brown, eds. Assessing the Vietnam War. Washington, DC: Pergamon-Brassey's International Defense Publishers, Inc., 1987.

McNeill, William H. The Rise of the West. Chicago: University of Chicago Press, 1963.

Millett, Allen R. and Williamson Murray. "Lessons of War." The National Interest (Winter 1988-1989): 83-95. Quoted in Air Force Manual 1-1.

Mitchell, William. Winged Defense. New York and London: G.P. Putman's Sons, 1925, 164. Quoted in Barry D. Watts. The Foundations of US Air Doctrine, The Problem of Friction in War. Maxwell Air Force Base, AL: Air University Press, December 1984.

Mooney, James D. "The Principles of Organization." in Papers on the Science of Administration, ed. Luther Gulick and L. Urwick, 89-98. New York: Institute of Public Administration, 1937.

Morgenthau, Hans J. Politics Among Nations. Revised by Kenneth W. Thompson. 6th ed. New York: Alfred A. Knopf, Inc., 1985.

Nelson, Harold D., ed. South Africa: A Country Study. Area Handbook Series. Washington, DC: U.S. Government Printing Office, 1981.

Norval, Morgan. Inside the ANC: The Evolution of a Terrorist Organization. Washington, DC: Selous Foundation Press, 1990.

Nossiter, Bernard. Washington Post, 3 February 1976.

Nowlin, David V. and Ronald J. Stupak. "War Termination and the Persian Gulf War: Getting-In, Getting-Out, and Getting It Right." American Society for Public Administration: SNSDA 11, no.1 (Summer 1991): 1-8.

Nye, Joseph S. Bound to Lead: The Changing Nature of American Power. New York: Basic Books, Inc., 1990.

Ott, J. Steven. Classic Readings in Organizational Behavior. Pacific Grove: Brooks/Cole Publishing Company, 1989.

Palmer, Dave Richard. Summons of the Trumpet: U.S.-Vietnam in Perspective. San Rafael, CA: Presidio Press, 1978, 110. Quoted in Harry G. Summers. On Strategy, A Critical Analysis of the Vietnam War, 127. New York: Dell Publishing Co., Inc., 1982.

Paret, Peter, ed. Makers of Modern Strategy From Machiavelli to the Nuclear Age. Princeton: Princeton University Press, 1986.

Patton, Michael Quinn. How to Use Qualitative Methods in Evaluation. Newbury Park, CA: Sage Publications, Inc., 1987.

_____. Qualitative Evaluation and Research Methods. 2nd ed. Newbury Park, CA: Sage Publications, Inc., 1990.

Porter, Bruce D. The USSR in Third World Conflicts. Cambridge, U.K.: Cambridge University Press, 1984.

Powell, Colin L. Interview by author, 10 May 1994, Pretoria, South Africa.

_____. News conference in Washington, DC, on 23 January, 1991. Quoted in Harry G. Summers. A Critical Analysis of the Gulf War. New York: Dell Publishing, 1992.

Preston, Richard A. and Sydney F. Wise. Men in Arms, A History of Warfare and Its Interrelationships with Western Society. 4th ed. New York: Holt, Rinehart and Winston, 1978.

Reardon, Steven L. The Evolution of American Strategic Doctrine. No. 4, SAIS Papers in International Affairs. Boulder: Westview Press, 1984.

Record, Jeffrey. Revising U.S. Military Strategy, Tailoring Means to Ends. Washington: Pergamon-Brassey's International Defense Publishers, 1984.

Republic of South Africa. Minister of Defence. White Paper on Defence and Armaments Supply—1986. Cape Town, S.A.: Naval Printing Press, 15 April 1986.

Republic of South Africa. Ministry of Defence. White Paper on Defence and Armaments Supply—1984. Cape Town, S.A.: Naval Printing Press, 1984.

Republic of South Africa. Parliament. Defence Act No. 44 of 1957. Statutes of the Republic of South Africa—Defence. Vol. 13 (1957).

Republic of South Africa. South African Army. Conventional Land Battle. Pretoria: South African Army.

Robbins, Stephen P. Essentials of Organizational Behavior. 2nd ed. Englewood Cliffs: Prentice Hall, 1988.

Sanders, Ralph and Edwin Timbers. Guide to the Analysis of Management Cases. Washington, DC: Fort Lesley J. McNair, 1983.

Sass, Brigadier W.P. (SADF, Ret.). 1993. Interview by author, 27 July, Pretoria, South Africa.

Satchwell, Kathy. "The Power to Defend: An Analysis of Various Aspects of the Defence Act." In War and Society: The Militarisation of South Africa, ed. Jacklyn Cock and Laurie Nathan, 40-50. Claremont, South Africa: David Philip, Publisher (Pty.) Ltd., 1986.

Savimbi, Jonas. Interview by Fred Bridgland, 4-5 July 1980, London, England. Cited in Fred Bridgland. Jonas Savimbi: A Key to Africa, 258. Johannesburg: Macmillan South Africa (Publishers) (Pty.) Ltd., 1986.

Schandler, Herbert Y. "America and Vietnam: The Failure of Strategy, 1964-67." in Vietnam as History, Ten Years After the Paris Peace Accords, ed. Peter Braestrup, 23-33. Washington, DC: University Press of America, 1984.

_____. The Unmaking of a President, Lyndon Johnson and Vietnam. Princeton, NJ: Princeton University Press, 1977.

Schwarzkopf, H. Norman. "Central Command Briefing." Briefing to the press in Riyadh, Saudi Arabia, on 27 February 1991. Cited in Military Review, LXXI, no. 9 (September 1991): 96-108.

_____. It Doesn't Take a Hero. New York: Bantam Books, Original Paperback, 1993.

Seabury, Paul and Angelo Codevilla. War, Ends & Means. New York: Basic Books, Inc., 1989.

Shafritz, Jay M. and J. Steven Ott. Classics of Organization Theory. 2nd ed. Chicago: The Dorsey Press, 1987.

Shuman, Howard E. and Walter R. Thomas, eds. The Constitution and National Security. Washington: National Defense University Press, 1990.

Shy, John. "Jomini." in Makers of Modern Strategy From Machiavelli to the Nuclear Age, ed. Peter Paret, 143-185. Princeton: Princeton University Press, 1986.

Simon, Herbert A. Administrative Behavior: A Study of Decision-Making in Administrative Organization. 3rd ed. New York: The Free Press, 1976.

Smith, Paul A., Jr. On Political War. Washington: National Defense University Press, 1989.

Snyder, Richard C., H.W. Bruck, and Burton Sapin, eds. Foreign Policy Decision-Making. New York: The Free Press of Glencoe, 1962.

Sparks, Allister. The Mind of South Africa. The New Hotfire Trust, 1990; First Ballantine Books edition, New York: Ballantine Books, 1991.

Staudenmaier, William O. "Conflict Termination in the Nuclear Era." In Conflict Termination and Military Strategy, Coercion, Persuasion, and War, ed. Stephen J. Cimbala and Keith A. Dunn, 15-32. Boulder: Westview Press, 1987.

Steenkamp, Willem. South Africa's Border War: 1966-1989. Gibraltar: Ashanti Publishing Limited, 1989.

Stockwell, John. In Search of Enemies: A CIA Story. New York: W.W. Norton & Company, 1978.

Stork, Joe and Martha Wenger. "From Rapid Deployment to Massive Deployment." In The Gulf War Reader, ed. Micah L. Sifry and Christopher Cerf, 34-39. New York: Random House, Inc., 1991.

Stupak, Ronald J. "Military Professionals and Civilian Careerists in the Department of Defense." Federal Manager's Quarterly (November 1988): 19-26.

Stupak, Ronald J. and Thomas C. Hone. "National Security and Domestic Policy-Making: The Similarities and the Critical Differences." International Journal of Public Administration 15, no. 7 (1992): 1441-48.

Summers, Harry G. "Lessons: A Soldier's View." in Vietnam as History, Ten Years After the Paris Peace Accords, ed. Peter

Braestrup, 109-114. Washington, DC: University Press of America, 1984.

_____. On Strategy, A Critical Analysis of the Vietnam War. New York: Dell Publishing Co., Inc., 1982.

Sun Tzu. The Art of War. Translated by Samuel B. Griffith. New York: Oxford University Press, 1971.

SWAPO Department of Information and Publicity. To Be Born a Nation: The Liberation Struggle for Namibia. London: Zed Press, 1981.

Tang, Truong Nhu. "The Myth of a Liberation." New York Review of Books, 21 October 1982, 31-36. Quoted in Harry G. Summers, Jr. "A Strategic Perception of the Vietnam War," Parameters 13, no. 2 (June 1983): 41-46.

Taylor, Frederick Winslow. The Principles of Scientific Management. Norwood, MA: The Plimpton Press, 1911; reprint, New York: Harper & Brothers Publishers, 1934.

Taylor, Maxwell D. The Uncertain Trumpet. Westport: Greenwood Press, 1960.

Thompson, James D. Organizations in Action. New York: McGraw-Hill Book Company, 1967.

Toth, James E. "Winning War and Peace." In The Soviet Challenge in the 1990s, ed. Stephen J. Cimbala, 167-183. New York: Praeger, 1989.

Tsouras, Peter and Elmo C. Wright, Jr. "The Ground War." In Military Lessons of the Gulf War, ed. Bruce W. Watson, 15-18. 2nd rev. ed. London: Greenhill Books, 1993.

Ungar, Sanford J. Africa: The People and Politics of an Emerging Continent. 3rd rev. ed. New York: Simon & Schuster, Inc., 1989.

Urwick, L. "Organization as a Technical Problem." in Papers on the Science of Administration, ed. Luther Gulick and L. Urwick, 47-88. New York: Institute of Public Administration, 1937.

U.S. Congress. Senate. Committee on Armed Services. Goldwater-Nichols Department of Defense Reorganization Act of 1986. Report no. 99-280, 99th Congress, 2nd Session, 1986.

U.S. Department of Defense. Air Force. Air Force Manual 1-1, Basic Aerospace Doctrine of the United States Air Force, vol. II. Washington, DC: U.S. Government Printing Office, March 1992.

U.S. Department of Defense. Army. FM 100-5, Operations. Washington, DC: U.S. Government Printing Office, May 1986.

U.S. Department of Defense. Conduct of the Persian Gulf Conflict: An Interim Report to Congress. Washington, DC: Department of Defense, July 1991.

U.S. Department of Defense. Conduct of the Persian Gulf War: Final Report to Congress. Washington, DC: Department of Defense, April 1992.

U.S. Department of Defense. Joint Chiefs of Staff. JCS Pub 2, Unified Action Armed Forces (UNAAF). Washington, DC: Department of Defense, The Joint Chiefs of Staff, December 1986.

U.S. Department of Defense. Joint Chiefs of Staff. JCS Pub 3-0 (Test Pub); Doctrine for Unified and Joint Operations. Washington, DC: Department of Defense. The Joint Chiefs of Staff, January 1990.

U.S. Department of Defense. Joint Staff. Joint Pub 0-1 (Proposed Final Pub); Basic National Defense Doctrine. Washington, DC: Department of Defense. Joint Staff, 7 May 1991.

U.S. Department of Defense. Joint Staff. Joint Pub 1, Joint Warfare of the US Armed Forces. Washington, DC: Department of Defense. The Joint Chiefs of Staff, 11 November 1991.

U.S. Department of Defense. Joint Staff. <u>Joint Pub 1-02; Department of Defense Dictionary of Military and Associated Terms</u>. Washington, DC: U.S. Government Printing Office, Department of Defense, Joint Staff, 1 December 1989.

U.S. Department of Defense. <u>Report of the Secretary of Defense Caspar W. Weinberger to the Congress on the FY 1987 Budget, FY 1988 Authorization Request and FY 1987-91 Defense Programs</u>, by Caspar W. Weinberger. Washington, DC: U.S. Government Printing Office, 1986, 273. Quoted in Stephen J. Cimbala and Keith A. Dunn, eds. <u>Conflict Termination and Military Strategy: Coercion, Persuasion, and War</u>. Boulder: Westview Press, 1987.

U.S. Department of Defense. <u>Report of the Secretary of Defense Frank C. Carlucci to the Congress on the FY 1990/FY 1991 Biennial Budget and FY 1990-94 Defense Programs</u>, by Frank C. Carlucci. Washington, DC: Department of Defense, January 17, 1989.

U.S. Department of State. "Regional Issues and U.S.-Soviet Relations," by Michael H. Armacost, Current Policy No. 1089, June 22, 1988, 5. Quoted by Peter Vanneman in <u>Soviet Strategy in Southern Africa</u>. Stanford: Hoover Institution Press, 1990.

U.S. News & World Report. <u>Triumph Without Victory, The History of the Persian Gulf War</u>. New York: Times Books, 1993.

U.S. Pacific Fleet. Headquarters of the Commander in Chief. <u>Outline Campaign Plan—Granite II</u>. Pearl Harbor, HI: HQ of the Commander in Chief, 3 June 1944.

Van der Waals, Willem S. (Brigadier, SADF, Ret.). <u>Portugal's War in Angola, 1961-1974</u>. Rivonia, South Africa: Ashanti Publishing (Pty.) Ltd., 1993.

_____. Interview by author, 9 July 1994, Pretoria, South Africa.

Vanneman, Peter. <u>Soviet Strategy in Southern Africa</u>. Stanford: Hoover Press, 1990.

Venter, Albert, ed. South African Government and Politics. Johannsburg: Southern Book Publishers (Pty.) Ltd., 1989.

Watson, Bruce W. and Bruce W. Watson, Jr. "The Iraqi Invasion of Kuwait." in Military Lessons of the Gulf War, ed. Bruce W. Watson, 15-18. 2nd rev. ed. London: Greenhill Books, 1993.

Weber, Max. Max Weber, Selections in Translation. Edited by W.G. Runciman. Cambridge: Cambridge University Press, 1978.

Weigley, Russell F. The American Way of War. New York: Macmillan Publishing Co., 1973; Bloomington, IN: Indiana University Press, 1977.

Wheeler, Michael O. Nuclear Weapons and the National Interest: The Early Years. Washington: National Defense University Press, 1989.

White House. National Security Strategy of the United States. Washington, DC: 1990.

White, Leonard. "The Meaning of Principles in Public Administration." In The Frontiers of Public Administration, ed. John M. Gaus, Leonard D. White, and Marshall E. Dimock, 13-25. Chicago: The University of Chicago Press, 1936; reissue, New York: Russell & Russell, 1967.

Wylie, J.C., Rear Admiral USN (Ret.). Military Strategy: A General Theory of Power Control. (New Brunswick, NJ: Rutgers University Press, 1967; reprint, Westport, CT: Greenwood Press, 1980.

Yin, Robert K. Case Study Research, Design and Methods. Vol. 5, Applied Social Research Methods Series. Newbury Park: Sage Publications, 1989.

Young, Tom. "Angola: Peace at Last." In South Africa at the Crossroads? ed. Larry Benjamin and Cristopher Gregory, 19-39. Rivonia, South Africa: Justified Press, 1992.

Index

Afghanistan, 1, 167

ANC, 88, 96, 134, 140, 141, 151, 168, 179

Angola, 1, 37, 84, 85, 88, 89, 91, 92, 94, 95, 127-40, 144, 152, 154-59, 161, 162, 164-69, 178, 179, 182-83, 185-90

Apartheid, 87, 88, 140

Botha, P.K., 88, 96, 97, 131, 133, 149, 153, 157, 166, 167

Botha, R.F., 96

Brezhnev, Leonid, 95, 96, 134

Burma, 1

Bush, George, 42, 69, 72, 94, 101, 103, 105, 106, 111, 124, 126, 129, 135, 146, 161, 163, 170, 176, 182

Castro, Fidel, 1, 134, 137, 159

China, 1, 4, 5, 44, 146

Chipenda, Daniel, 138

CIA, 130, 133

Clark Amendment, 159-62

Cold War, 1, 78, 97, 100, 102

Colonialism, 136, 138

Communism, 88, 95, 96, 129, 130, 133, 134, 140, 151, 152, 167, 168

Cuba, 1, 15, 88, 95, 130-9, 141, 144-7, 149-53, 159, 161, 165-7, 182-185

Cuito Cuanavale, 151-53, 159-62, 179, 182-85

Decision Making, 3, 5, 11-20, 23-26, 41, 43, 45, 47, 60, 67, 72, 73, 60, 67, 72, 73, 78-81, 84, 87-89, 96, 97, 169, 171, 172, 175, 176, 184, 188

Defense Department, 17, 34, 35, 72, 73, 75, 78, 80, 116, 121, 125

Desert Shield, 103, 104, 109, 112, 120, 121

Desert Storm, 112, 114, 119-21, 174, 176

Eisenhower, Dwight, 46

FNLA, 130-33, 135, 137, 139, 146

Gorbachev, Mikhail, 166

Great Britain, 44, 100, 105

Greece, 1, 84

Groupthink, 16, 17

Guerrilla Warfare, 1, 95, 129, 131, 133, 144, 145, 147, 151, 152

Gulf War, 2, 7, 29, 55, 62, 78, 97, 102, 112, 116, 121, 124, 125, 169, 170, 172, 176-78, 186, 187, 190

Hussein, Saddam, 100, 104, 105, 107, 110, 122-126, 174-177, 190

Imperialism, 138, 140

Iran, 73, 100, 116, 126, 190

Iraq, 57, 72, 97, 100-4, 106, 107, 109-12, 114, 116, 119-25

Israel, 190

Japan, 47, 62

Johnson, Lyndon, 3-7

Joint Chiefs of Staff, 5, 7, 60, 61, 112

Korean War, 2, 48, 63, 105

Kuwait, 57, 72, 97, 100, 101, 103, 104, 106, 107, 110, 111, 113, 114, 116, 119, 120-24, 174-76, 190

Lusaka Accords, 148-150, 165

National Party, 86, 87, 140

Philippines, 1

Portugal, 129, 130, 138, 146

Principles of War, 1, 30, 37-44, 72, 83, 94, 125, 169, 177, 178, 185-90

Public Administration, 11, 13, 18-20, 22-24

Powell, Colin, 171, 177

Reagan, Ronald, 101, 166

Roberto, Holden, 138, 139, 146

Russia, 2, 5, 102, 133, 135, 136, 149, 152, 182, 183

Saudi Arabia, 100, 101, 103-5, 107, 114, 119, 121-23, 172, 174, 176

Savimbi, Jonas, 95, 132, 135, 137, 139, 145-47, 150, 152, 167, 168

Schwarzkopf, Norman, 7, 103, 107, 171, 172, 175, 176

South Africa, 44, 84-90, 92-97, 129-34, 137-41, 143-57, 161, 163-68, 179, 183, 185

South West Africa (Namibia), 95, 130, 133, 134, 138, 143, 144-49, 152, 156, 165, 166, 179, 181

Soviet Union, 2, 4, 44, 71, 95, 102, 106, 130, 131, 133-35, 137-39, 141, 166, 167

Stalin, Joseph, 140

Sun Tzu, 18, 29, 30, 31, 49, 53, 56, 57, 63, 83, 94

SWAPO, 134, 144-45, 147-49, 151, 152, 165, 179, 181

Systems Approach, 24-26

UNITA, 130-35, 137-40, 145-48, 150-54, 157-59, 161-65, 167, 179, 182-84

United Nations, 96, 104, 106, 133, 143, 144, 166, 167

United States, 1-3, 5, 8, 29, 36, 44, 48, 49, 55, 58, 62, 63, 67, 69-71, 83, 90, 97, 100-04, 126, 130, 132-34, 137, 145-46, 150, 177, 167, 174

Unity of Command, 6, 19, 21, 44, 54, 55, 171, 172, 179

Vietnam, 2-8, 17, 47-49, 51, 55, 58, 63, 73, 83, 102, 112, 113, 126, 137, 178

World War I, 29, 44, 46, 51, 54, 100, 143

World War II, 1, 5, 13, 22, 29, 46, 51, 55, 58, 58, 62, 63, 72, 73, 79, 83, 84, 97, 102, 140, 171